More Molecules of Murder

More Molecules of Murder

John Emsley
Email: JohnEmsley38@aol.com

Print ISBN: 978-1-78801-103-7
EPUB eISBN: 978-1-78801-238-6

A catalogue record for this book is available from the British Library

The Royal Society of Chemistry is a charity, registered in England and Wales, Number 207890, and a company incorporated in England by Royal Charter (Registered No. RC000524), registered office: Burlington House, Piccadilly, London W1J 0BA, UK, Telephone: +44 (0) 207 4378 6556.

Visit our website at www.rsc.org/books

Printed and bound in the United States of America

Acknowledgements

I should like to thank the following people for their help in writing this book. Some of them provided me with information I would not otherwise have had, but most of them checked the contents of various chapters, where their expert knowledge enabled them to say whether the science was correct.

In alphabetical order, they are as follows:

Professor Anil Aggrawal of the Maulana Azad Medical College, New Delhi, India, and Editor-in-Chief of the *Internet Journal of Forensic Medicine and Toxicology*.

Professor Robert Chilcott, a toxicologist based at the University of Hertfordshire.

Dr Simon Cotton of the School of Chemistry at Birmingham University.

The late Alistair Crawford MD of Ampthill.

Dr Catherine J. Duckett, Lecturer in Analytical and Forensic Chemistry at Keele University.

Helen Glyn-Davies RD, registered dietitian working with the NHS.

Dr Fritof Koerber of Koerber Education Services, Bristol, and senior lecturer at the University of the West of England.

Dr Caroline Lockley of Hampshire.

Dr John Lockley of Ampthill.

More Molecules of Murder
By John Emsley

Published by the Royal Society of Chemistry, www.rsc.org

Dr Gillian Marsh, toxicologist of Northumberland.
Dr Michael Utidjian of Wayne, New Jersey, and former Corporate Medical Director of the former American Cyanamid Co.

I also prevailed upon my wife, Joan Emsley, and my friends Professor Steve Ley of the Department of Chemistry, University of Cambridge, and his wife Rose, to read the complete text, and their comments proved invaluable. I would also like to thank Kathryn Duncan for editing the text and Susannah May for her support. I would like to think this book is now free of errors, but if there are still some to be found, then I accept full responsibility and would welcome feedback from readers.

Biography

Dr John Emsley is best known for his series of highly readable popular science books about everyday chemistry, some of which have run into multiple editions and printings in the UK and been translated into several other languages. He has also published in national newspapers and magazines, and he has written chemistry text books and booklets for industry. John has carved an impressive career in popular science writing and broadcasting over the past 25 years, emphasising the benefits of chemistry, and the agrochemical, pharmaceutical, and chemical industries. His books include:

The Chemistry of Phosphorus (with C. D. Hall), 1976
The Elements (1st, 1987, also 2nd and 3rd editions)
The Consumer's Good Chemical Guide, 1994 (Science Book Award Winner)
Molecules at an Exhibition, 1998
Was it Something You Ate? (with Peter Fell) 1999
The Shocking History of Phosphorus, 2000
Nature's Building Blocks, 2001
Vanity, Vitality, Virility, 2004
Elements of Murder, 2005
Better Looking, Better Living, Better Loving, 2007
Molecules of Murder, 2008
A Healthy, Wealthy, Sustainable World, 2010

More Molecules of Murder
By John Emsley

Published by the Royal Society of Chemistry, www.rsc.org

Nature's Building Blocks (2nd edition), 2012
Islington Green, 2012
Sweet Dreams, 2013
The Newsletter, 2013
Chemystery, 2013
Chemistry at Home, 2015
Elements of Fortune, 2016
Chemhistory, 2017

Contents

More Molecules of Murder
By John Emsley

Published by the Royal Society of Chemistry, www.rsc.org

Introduction

My previous book, *Molecules of Murder*, was divided into two parts. The first consisted of five chapters dealing with natural materials that have been used to murder, these being ricin, hyoscine, atropine, diamorphine, and adrenaline; the second group of five chapters dealt with man-made chemicals, namely chloroform, carbon monoxide, sodium cyanide, paraquat, and polonium chloride. I have constructed this present book along similar lines. Part I consists of seven chapters which are about man-made chemicals, namely ethylene glycol, acrylamide, difenacoum, oxalic acid, temazepam, potassium chloride, and tetramethylenedisulfotetramine, while Part II has seven chapters dealing with natural toxins, namely gelsemine, strychnine, digitalis, curare, aconitine, cantharidin, and hemlock.

More Molecules of Murder and its predecessor *Molecules of Murder* differ from most true-crime murder books and novels in that they focus not on the murderer, the victim, the reason for the crime, or the detectives who solve the case, but on the poisonous agent that was used. At the start of every chapter, I explain the chemistry and toxicity of the chosen agent before analysing various murders and deaths which have been caused by it.

Fictional crime series grace our TV screens most weeks, and while forensic investigations are now referred to, those who appear are generally pathologists with scenes invariably set in

---

More Molecules of Murder
By John Emsley

Published by the Royal Society of Chemistry, www.rsc.org

mortuaries. Rarely do chemists get a look in. It was not always so; in the tales of Sherlock Holmes, it was chemistry that often took centre stage.

Holmes was the best detective who never lived. He was created by Arthur Conan Doyle and he first appeared in *A Study in Scarlet*, published in 1887. Holmes would often resort to chemical analysis in solving crimes and Doyle, although not a chemist himself, generally got things correct thanks to his close association with the chemistry department of University College London. In fact, Holmes was well ahead of his time with regard to forensic analysis and his apparent success may well have encouraged the setting up of laboratories to promote this aspect of solving crimes, which today we take for granted.

Another writer who relied on chemical knowledge and whose works became equally famous was Agatha Christie. She was the leading crime novelist of the last century, and she used death by poison in most of her books – and she almost always got the toxicology correct. She became interested in the subject when, as a young woman, she did voluntary work in the hospital in Torquay in World War I. There, she trained as the apothecary's assistant in the dispensary, and she became skilled in the making of pills, potions, and tonics.

Christie's main desire, however, was to become a novelist, and in her spare time, she wrote articles and eventually books, of which *The Mysterious Affair at Styles* became her first success in 1919. In that book, she used strychnine to dispatch her victim.[†] And so began a career that was to give her worldwide fame as a crime novelist, writing more than 60 novels and selling over two billion copies. She also wrote plays, of which the best known is *The Mousetrap.* This is the play with the longest-ever run, having been performed more than 25 000 times during 60 years at the St Martin's theatre in London.

Two hundred years ago, life for most people was hard; food might be in short supply, many illnesses could not be cured, pain was something you had to live with, homes were ill equipped and possibly dirty, clothes were made only from natural

[†]Kathryn Harkup's 2015 book *A is for Arsenic* is a wonderful analysis of Christie's novels from the point of view of the poisons used and shows that, although the novelist was not a trained chemist, she knew what she was writing about.

fibres and were not easily cleaned. It was chemistry and its related industries that changed all this and so transformed everyday living, with chemicals that resulted in abundant food, better health, stronger materials, softer fabrics, brighter colours, cleaner homes, unbreakable plastics, and less pain.

But, of course, it also created some nasty chemicals that were as poisonous as those which Nature creates. However, it is virtually impossible for members of the public to access them. The result has also been a decline in murder by poison, and this is reflected in today's crime novels. Nevertheless, poisonings still do occur but the US website http://www.poison.org/poison-statistics-national shows that the vast majority of these are accidental (often involving a child unintentionally drinking a household chemical). There are still a few deliberate cases of wilful poisonings, although the chances of these being successful are now much rarer, thanks to rapid analysis to identify the poison and medical attention to deal with its action.

In theory, poison is still a way in which someone can carry out a murder and produce symptoms that would mimic those of a natural illness so that death would be attributed to that, and no further action would then need to be taken in terms of forensic investigation. Thankfully, they generally fail, and there are modern cases covered in this book in which some murders were successful and some were not, and all are revealing.

HOW DO WE DEFINE A POISON AND HOW IS IT TO BE DETECTED?

The three most important things which determine whether a chemical can be described as a poison are (1) its effect on the human body, (2) its speed of action in the body, (3) the effect it has in small doses. Most people will be aware of these, and a truly deadly poison like cyanide meets all three of these requirements, hence its use in suicide capsules.

Poison given with malicious intent is generally disguised as food or drink and so is absorbed through the stomach wall or the intestines, which are particularly well supplied with blood vessels. Then it will be examined by the liver, which may realise it is not something the body's metabolism needs and may begin to change it chemically prior to its removal *via* the urine or faeces.

Were it to be injected directly into the bloodstream, then its circulation throughout the body might well ensure a rapid death, as we shall see.

Today, if there is the least suspicion that someone has been poisoned, it will be proved. Forensic chemists have tests and equipment that can detect and measure the amount of something that is present in blood or tissue, even when this is as little as a trillionth of a gram. So how does a forensic chemist do this? An obvious way is to use the method known as high performance liquid chromatography coupled to mass spectrometry (HPLC–MS) and compare the findings with the libraries of known poisons. Chromatography is a way of separating complex mixtures and mass spectrometry is a way of identifying even the smallest amount of material.

Forensic scientists no longer require the large amounts of body tissue which used to be needed for analysis; now, a few milligrams or even micrograms will suffice. Today, there are journals devoted to toxicology, such is the interest in this subject, and toxicology degrees are on offer at several universities.

Usually, when someone has died under suspicious circumstances, then blood samples are taken, and these will generally reveal whether they have been poisoned. Urine samples and stomach contents can also be analysed. However, when it comes to analysing a body that has been dead for some time, then other parts of the corpse may provide the information. Hair will contain a record of chemicals to which the body has been exposed. Tissue, likewise, can retain a record for a long time after a person has died.

When the cause of death is suspicious, then the toxicologist must begin a complex and systematic analysis, and the samples most likely to yield the answer are (in order of their ability to provide the necessary evidence): blood, urine, liver, bile, vitreous humour (from the eyes), cerebrospinal fluid (from the brain), brain tissue, lungs, hair, and nail clippings. Urine is less important than blood because the amount in the urine can vary widely depending on what the person had been drinking, the degree of hydration of the body, and the condition of the kidneys. When it comes to analysing bodies that are partly decomposed, then samples of muscle, skin, bone, hair, and even the maggots on the body are collected for analysis.

Toxicologists are now more likely to be concerned with analysing for recreational drugs, alcohol, or drugs like steroids which sports people use to boost their performance. Rarely would such a person be called upon to analyse for poisons of the kind discussed in this book and which have been given deliberately to kill. However, when he or she does analyse for them, then their evidence is likely to be incontrovertible; although, in one famous trial in New Zealand, the evidence was questioned because the process of analysis was so new its reliability was challenged, as you will read.

In an age when it is almost impossible for the average person to obtain toxic chemicals, some have resorted to using a medicine that can be fatal in excess, and one such common medicine is temazepam, as we shall see.

AN ANCIENT HISTORY OF POISONS

Poisons have been known for thousands of years; indeed, there is evidence that hunters in Kenya used poisons to kill their prey more than 15 000 years ago.[‡] No doubt this information about toxic agents was discovered accidentally by eating different plants, noting their effects, and maybe sometimes suffering the consequences. The poisonous extract we refer to as curare was certainly long-known to the Amazon tribes, and they even graded its strength as one-tree, two-trees, or three-trees – this being the number of trees a monkey could jump before falling victim to a poisoned arrow or dart. A three-tree curare was thought too inferior for hunting purposes. Nor was it just in South America that this type of weapon was used. The book of Job in the *Bible* mentions poisoned arrows:

> The arrows of the Almighty are in me, my spirit drinks in their poison;
> God's terrors are marshalled against me.

Poisons were certainly known in ancient Assyria, ancient India, and ancient Egypt. Clay tablets from Assyria show knowledge of

[‡]A. Aggrawal, 'A History of Toxicology,' *Encyclopaedia of Forensic and Legal Medicine*, Elsevier, London, 2005, vol. 2, pp. 525–538.

both dangerous metals and plant poisons. In Egypt, the so-called *Ebers Papyrus* was unearthed in Thebes by Georg Ebers in 1872. This was written around 1500 BC and is a list of 800 recipes for potions, some of which we can still recognise. In ancient Greece, a pupil of Aristotle called Theophrastus, who lived from 371 to 287 BC, wrote *De Historia Planarum*, in which he mentioned poisonous plants. He was aware of the execution of the philosopher Socrates in 399 BC who was forced to drink hemlock, which will be discussed in Chapter 14. In the second century AD, the Greek physician Galen wrote a book which gave antidotes for various poisons.

Poisons were well known in the Roman Empire, where they even disposed of Emperors – namely Augustus in 14 AD, Claudius in 54 AD, and Trajan in 117 AD. A Greek physician, Dioscorides (who was born around 40 AD and died around 90 AD), wrote a classification of poisons and listed them as animal poisons (toads, snakes, *etc.*), plant poisons (opium, mandrake hemlock, aconite, *etc.*) or mineral poisons (arsenic, mercury, lead, *etc.*).

In 331 BC, a group of women in Rome were arrested and put on trial accused of poisoning people. They were found guilty and then forced to consume the suspect liquid found in their home. They died. A similar group of women, many of noble birth, were rounded up in 200 BC and executed for poisoning people. In 183 BC, the great Carthaginian general Hannibal committed suicide by drinking cyanide, which can be obtained from the stones of peaches.

In the centuries following Charlemagne, it is suspected that several of his successors were poisoned, and five popes are thought to have suffered a similar fate. For example, Pope Clement VII may have been murdered by being fed poisonous mushrooms in 1534. King John of England was believed to have been poisoned in 1216 with toad toxin. And so it went on.[§]

One of the most prolific poisoners in history was Catherine Deshayes, better known as La Voisine (The Neighbour). She is believed to have sold poisons to thousands of people who wished to dispose of husbands and relatives. She was burnt at the stake in 1680.

[§]Several famous historical cases are included in my book *The Elements of Murder*, OUP, 2006.

Nor is the New World without its attempted poisoning of leaders. In 1776, Thomas Hickey tried unsuccessfully to assassinate George Washington and was executed for treason.

Antidotes for poisons were sought in the ancient world, and most famous for his research in this area was King Mithridates VI of Pontus in Asia Minor, who lived from 135 to 63 BC. His method of testing was to use condemned prisoners, who were fed the poison and then given a supposed antidote. This way, he discovered several useful substances and devised a universal antidote, which included more than 50 ingredients. This he took daily as a form of protection, and it appeared to work because, when he was defeated by the Romans, he took poison, which didn't do the job. (He was eventually stabbed to death by ordering one of his guards to kill him.)

A useful antidote material discovered in ancient times was terra sigillata, an absorbent clay from the island of Lemnos. It was used until early in the last century, and it might well have saved the lives of some individuals by absorbing poison in the gut and carrying it safely out of the body.

Testing for poisons in corpses only became possible with the advances of chemical analysis and even then it was not easily done. Testing for arsenic became possible thanks to the Marsh Test, developed in 1836. However, plant-based poisons presented difficulties. The first forensic scientist to devise a method of testing for such poisons was the Belgian Jean Servais Stas (1813–1891), who successfully proved that Hippolyte Visart de Bocarmé had used nicotine to murder Gustave Fougnies for his money. de Bocarmé was executed by guillotine in July 1851.

Slowly, the advances in chemical separation and analysis made it possible to provide courts with scientific evidence of poisoning in criminal cases. This was not without its own problems to begin with because of the need to explain to the jury the methods used and the amounts detected. When such evidence was based on a new method of analysis, then it was easy for defence lawyers successfully to question its reliability. Nor has this changed, as we shall see.

Things to Bear in Mind

More Molecules of Murder is meant to be read by anybody, not just chemists, and yet I have to use words and measurements that chemists are familiar with which may be unfamiliar to the average reader. Where these are key to a story, they will be highlighted in **bold**, indicating that there is a fuller explanation in the Glossary.

When it comes to measurements, I use the metric system of grams, litres, and metres, rather than the Imperial system of ounces, pints, and feet. For the amounts of toxin that are detected in the blood of a person suspected of being poisoned, then the units are tiny and expressed in terms of milligrams (mg), which are a thousandth part of a gram, and micrograms (μg), which are a millionth part of a gram. Most toxicological data will be expressed in terms of how much there is in a litre of blood or a kilogram of tissue. For liquids like blood, it is sometimes reported in terms of micrograms per millilitre, or milligrams per litre, but these numbers are the same.

Alternatively, and more generally, the amounts can be expressed in terms of parts per million (ppm), which is one milligram per litre, or one microgram per millilitre. Even smaller concentrations can be reported as parts per billion (ppb), the latter being equivalent to a microgram in a litre. To give an idea

More Molecules of Murder
By John Emsley

Published by the Royal Society of Chemistry, www.rsc.org

of how small this last quantity is: 1 ppb is like one second in 30 years. It is even possible for modern analytical techniques to measure in parts per trillion (ppt), which is like one second in 30 000 years. Finding and measuring such tiny amounts is a tribute to the way forensic chemistry has developed in recent years.

Part I
Man-made Molecules

CHAPTER 1

Ethylene Glycol for Antifreeze and Loved Ones

A word in **bold** *indicates that further information can be found in the Glossary. Only the first time the word appears in a chapter will it be so indicated.*

Ethylene glycol may be poisonous but it is not a chemical we need worry about, even though we can easily purchase it to use as antifreeze. Indeed, most vehicles contain this chemical and it also has industrial applications. Most of the ethylene glycol that is produced industrially is used to make fibres and plastics. However, it is not entirely without risks, and this sweet-tasting liquid results in many accidental poisonings worldwide every year, especially those involving children who are attracted by its sweet taste. In the USA, there are about 6000 such cases reported annually, of which 20 are fatal. Some desperate individuals resort to drinking this chemical to commit suicide – but it's a slow death. To prevent ethylene glycol from being consumed by children, the bittering agent Bitrex is sometimes added, and in certain states of the USA this is mandatory, such as Oregon where it was introduced as a safety measure in 1991. Sixteen other states also have this requirement.

Sometimes ethylene glycol is used fraudulently. In the 1980s, unscrupulous Austrian vintners added ethylene glycol to cheap

More Molecules of Murder
By John Emsley

Published by the Royal Society of Chemistry, www.rsc.org

wines so they could pass them off as high quality and so charge a higher price. So why were their customers not affected by this chemical? Admittedly the amounts in wine were small, but in any case, the alcohol in the wine was acting as an antidote, something which would-be poisoners may not be aware of, as we shall discover.

Ethylene glycol is not as deadly as some of the other molecules in this book, but it can be used to murder someone if it is done carefully, and would-be murderers can discover its potential as a poison by searching the internet. There are several notable examples of its misuse, although often the intended victim has survived and lived to see their persecutor imprisoned. Sometimes, even seemingly respectable people such as doctors have resorted to using it.

1.1 ETHYLENE GLYCOL

This chemical finds its way directly into our lives in addition to being used as antifreeze. It is to be found in products such as shoe polish, tiling grout, gloss paints, and Tip-Ex correction fluid. It prevents, or stops, these products from drying out too quickly. However, the biggest use of this chemical is in our cars in engine coolants and windscreen washes (Figure 1.1). A typical wash will be 25% antifreeze with 75% water, or even a 50 : 50 mixture for extremely cold conditions when the temperature might fall to as low as −25 °C. Ethylene glycol is also extensively used in cold climates to remove the ice that can accumulate on

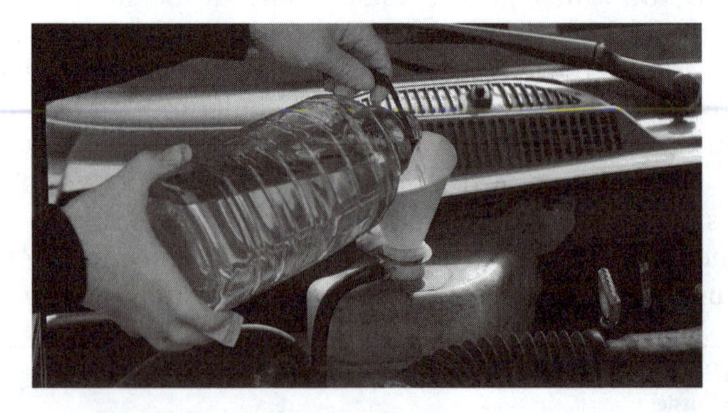

Figure 1.1 Adding coolant with ethylene glycol or antifreeze to an engine. © SEASTOCK/Shutterstock.

aircraft wings. However, this has caused local environmental issues because it gets washed on to land or into rivers.

Industrially, ethylene glycol is produced on a massive scale (18 million tonnes a year), most going to make polyethylene terephthalate, aka PET, which is used as a fibre for clothing and as a clear plastic for bottles. PET was discovered in Manchester, England, in 1941 by two chemists Rex Whinfield and James Dickson. They heated ethylene glycol with dimethyl terephthalate at 200 °C and noticed that a sticky material was produced which could be easily drawn into fibres, and that these were strong and unaffected by boiling water. When blended with cotton they produced a fabric which was comfortable to wear, did not crease, and was easy to launder. Polyester and polyester cotton now account for more than 50% of the clothes we buy.

In the 1960s, a new use for PET was found in the form of transparent plastic bottles. They save energy, because they require only 45% of that needed to make a glass bottle and they weigh much less. They save on transport, because a delivery truck can carry 60% more of such containers. And, unlike glass, they do not break into sharp fragments. While most such bottles are sold as non-returnable, it is possible to make them strong enough to be reused. Alternatively, PET bottles can be recycled into different products by being melted down and turned into plastic that is suitable for other types of packaging, or into polyester fibre and turned into such products as carpets, duvets, anoraks, bristles for paint brushes, and felt for tennis balls.

1.2 THE TOXIC NATURE OF ETHYLENE GLYCOL

The body has no use for ethylene glycol and so it changes it chemically as a way of disposing of it, and to do this it attacks it with the enzyme called alcohol dehydrogenase. This first metabolises it to **glycolaldehyde**, then to **glycolic acid**, glyoxylic acid, and finally to **oxalic acid**. These are what make ethylene glycol so dangerous because these acids are rather nasty chemicals which the body must dispose of, but this takes time. The most toxic of these by-products is oxalic acid, and this is the poison discussed in the next chapter.

What constitutes a deadly dose of ethylene glycol for an adult? If this is consumed as a 50:50 liquid with water then around

200 mL might well prove fatal. Symptoms of poisoning will begin within an hour or so and will result in vomiting, while the person affected may behave as if they are drunk. Later, more serious changes are occurring in the body as the toxins are produced which affect the heart and the breathing. Then, after a day or two, the kidneys begin to fail and the person may by then be in a coma or have suffered a seizure.

There are antidotes to ethylene glycol and these are alcohol itself (ethanol) or **fomepizole**, which is rather expensive although it works better. These block the enzymes that are responsible for converting ethylene glycol to the more dangerous toxins. Alcohol as an antidote can be taken in the form of neat gin, vodka or whisky. What the alcohol does is take priority with the enzyme thereby giving the body more time to dispose of the ethylene glycol.

The toxicity of ethylene glycol was unsuspected in the 1930s and this was to have unfortunate consequences in the USA.

The first antibiotic drug, **sulfanilamide**, was discovered in 1932 by a German chemist, Gerhard Domagk, who worked for IG Farben on new dyes. When he tested sulfanilamide on mice, some of which were infected with deadly streptococci bacteria, they recovered. So he tried it on his sick daughter when she had an infection and she got better almost immediately. This was the first antibiotic, and it was marketed as Prontosil. It was soon being used around the world and Domagk was awarded the Nobel Prize for physiology and medicine in 1939 for his work.

Then, in 1937, an American company, S. E. Massengill, began to produce their version of the new drug, and their chief chemist, Harold Watkins, thought the best way to dispense this medication was as a liquid, which he named Elixir of Sulfanilamide. The solvent he chose was ethylene glycol, and he did this because he found that sulfanilamide was not soluble enough in water to make a useful medicine. However, it was soluble in ethylene glycol and that's how it was dispensed, and to make it more palatable it was flavoured with raspberry extract.

Soon doctors in the USA were prescribing the elixir, but it was not long before they began to report to the American Medical Association that some patients, mostly children, who were given it died unexpectedly. There was a product re-call but not before

107 had died, and then it fell to a Dr Frances Kelsey to research the problem.[†] She discovered that it was not the drug which was causing these deaths but the ethylene glycol. The end result was a change in the law regarding testing, and to this day, the US Food and Drug Administration (FDA) has to approve all such products before they can be prescribed. The man responsible for the deadly medicine was Harold Watkins, and he was arrested but committed suicide before he could be brought to trial.

Still the use of ethylene glycol in medicines continues. In 2009 in Nigeria, a teething mixture for babies called My Pikin Teething Mixture caused the deaths of more than 80 babies because it used ethylene glycol as a solvent for paracetamol. Children who were given it suffered vomiting, diarrhoea, and convulsions before dying. This outbreak was a repeat of a similar incident in 1990 when 116 people died as a result of a cough syrup made with ethylene glycol.

1.3 MURDER, WALTHAM, MASSACHUSETTS, 2004

Sports drinks are designed to replace the glucose and minerals that your body needs when you engage in strenuous exercise and sweat a lot. In the USA, the popular drink for this is Gatorade, a bottle of which contains 56 grams of carbohydrate, 4.5 grams of sodium, and 120 milligrams of potassium. One sport-loving woman regularly drank Gatorade and it was to this that her husband added another ingredient: ethylene glycol. It was the life insurance on his wife Julie that motivated 31-year-old James Keown to murder her with antifreeze in 2004.

Julie was 31 years old and they lived near Kansas City, Missouri, where Julie's parents were soybean farmers. She worked as a nurse and James worked for the Learning Exchange as a web-designer. Then, in 2003, he asked if he could enrol on a master's degree course at the Harvard Business School. His employers were happy for him to do this and he persuaded them that he could continue working for them from his new home on the east coast. They agreed, and the Keowns moved in January

[†]She was also the person who prevented the morning sickness drug thalidomide from being marketed in the USA in 1962. She lived to be 101.

2004 to Waltham, a town of 60 000 inhabitants located about 11 miles from Boston. But things were not as they appeared.

James also continued to broadcast a regular slot on a Missouri radio station but his days as a radio commentator were numbered. He was cheating his employer and was misusing a website design that they had asked him to develop for the Learning Exchange. When they tried to get in contact with him *via* the Harvard Business School, they discovered he was not registered with them. What, in fact, James had done was register for a course about the internet at the Harvard Extension School but he failed his exams. His deception had been uncovered and so the Learning Exchange sacked him.

Soon James was borrowing heavily to keep up the pretence of being employed and eventually was $34 000 in debt, and then his beloved Jaguar car was repossessed. However, he knew he would soon be exposed as bankrupt so he decided to murder Julie for her life insurance, and he planned to do this by putting ethylene glycol in her Gatorade.

Julie was first hospitalised in August 2004 after her speech became slurred and she stumbled when she walked. The doctors who treated her diagnosed a kidney disease and she recovered and was allowed home. Then James gave her a bigger dose of ethylene glycol on 4 September. She was admitted to hospital again and was clearly struggling with a mysterious illness which caused vomiting and slurred speech and which seemed to be related to her kidneys malfunctioning. Meanwhile, James played the part of a devoted and caring husband, always pressing her to drink her Gatorade to replace her minerals. The hospital doctors eventually discovered that she had been poisoned by ethylene glycol, but despite their efforts to save her, Julie died on 8 September. Before any action could be taken against him, James returned to the Midwest and went to live in Jefferson City, Missouri, a small town of 45 000 inhabitants and considered to be one of the prettiest towns in the USA.

When Julie's parents learned of the cause of her illness, they confronted James and he said that he thought she had found a soft-drink bottle in which someone had been storing antifreeze, and that she had drunk it. It confirmed their suspicions that James had poisoned her and they went to the police. Unaware of this, James continued to believe that his plan had worked and he told a friend how he intended spending the money he would get from Julie's life insurance of $250 000. He said he planned to buy a new BMW car and even thought of building a new house. However, his days of freedom were numbered, and he was arrested in 2005 and charged with the murder of his wife. Nor did it help his case when police examined his computer and discovered that he had searched the internet using the phrase 'ethylene glycol death human' a few days before Julie was first admitted to hospital.

James was eventually found guilty by a jury and sentenced to life imprisonment without parole.

1.4 ATTEMPTED MURDER, STOKE-ON-TRENT, 2005

One advantage of ethylene glycol as a murder weapon is that it tastes rather nice, with a sweetish flavour, and it isn't difficult to disguise it by putting it in other drinks. However, a would-be murderer was unaware that the alcohol in a drink could neutralise the effect of the poison. Nevertheless, she left her husband blind and deaf, although the deafness has been partly corrected by new technology.

When 28-year-old Kate Knight decided to kill her 38-year-old husband Lee in 2005 she added ethylene glycol to red wine.[‡] They had married when she was only 19 years old and she now wished to be rid of him. She saw poisoning her husband as a way out of an intolerable situation caused by her reckless spending and lack of income. Because she knew nothing about ethylene glycol's toxicity, she added only small amounts to begin with and, while this made Lee ill, he soon recovered. She realised she would have to use a bigger dose.

Meanwhile, she would lavish loving care on her husband and show concern for his welfare. He even believed his wife had a job working at a call centre, but she was living on credit and eventually owed £17 000, having taken out a loan of £10 000 in October 2003 and a further one of £7000 in March 2004. Her financial worries would be solved if Lee died because he had life insurance of £130 000 paid for by his employer JCB,

---

[‡]Red wine might seem to be an ideal medium because the tannins it contains mask the sweetness of the ethylene glycol.

manufacturer of construction equipment. He worked at their Uttoxeter plant, about 15 miles from where they lived in Stoke-on-Trent.

Kate played the role of a caring wife very successfully; greeting him warmly when he came home from work and making sure he had red wine for his evening meal. On their wedding anniversary in April 2005, Kate gave Lee a particularly large dose of ethylene glycol to accompany an Indian takeaway which she had ordered, knowing that he was particularly fond of such dishes. Lee became so ill after this particular meal that he rang for an ambulance and was taken to hospital in Stoke-on-Trent where he started to recover.

This was something of a setback for Kate's plan, but she was not yet done with poisoning Lee. When she visited him in hospital the next day, she took him a litre bottle of flavoured water that she had also poisoned and which she insisted he should drink. Lee now became delirious and the doctors put him on dialysis in order to keep him alive. Eventually, he suffered a massive haemorrhage during which he lost six pints of blood. He lay in a coma for two months before he finally regained consciousness, and yet the true cause of his condition was still not recognised. Kate now said she was unable to visit him because she had to look after their son Jack. During this time, their car and other items were repossessed because of unpaid debts. Her attempt to murder Lee had failed, although her husband had been on the point of dying and was now blind and deaf.

So what gave Kate away? In fact, she had told a neighbour friend that she was going to poison Lee because she was no longer in love with him. The neighbour simply thought of these remarks as mere fantasy, but she told a cousin of Lee's what Kate had said, and the cousin told Lee's parents and they told the doctors at the hospital. At last they knew what they were dealing with and informed the police, who went round to Kate's home. There they discovered an almost empty bottle of antifreeze, a computer which had been used to search for poisons, and correspondence showing the mortgage had increased by £30 000 while Lee had been in hospital.

The jury at Stafford Crown Court took eight hours to convict Kate of attempted murder following a three-week trial in January

2008. She was sentenced to 30 years in jail and told that she would have to serve at least 15 years before being eligible for parole.

1.5 ATTEMPTED MURDER, HOUSTON, TEXAS, 2013

On 29 September 2014, Ana Gonzalez-Angulo was found guilty of poisoning her one-time lover George Blumenschein because he refused to give up his long-time partner and move in with her. Both Ana and George were doctors researching cancer at the MD Anderson Cancer Center, which is part of the University of Texas in Houston, USA. This is one of the world's leading institutes for this kind of research.

In fact, George had suspected his mistress of trying to poison him but couldn't believe it, so he did not report her to the authorities. However, when he was seriously poisoned in January 2013, and had to be treated for his condition, he made charges against her and Ana was arrested.

The two had started a relationship when they went on a business trip to Stockholm, although it got no further than oral sex – or so Blumenschein maintained at the trial. The affair with Ana had been going on for 18 months before Ana realised that he would never leave his partner for her, and so she decided that if she could not have him then no one could. She decided to murder him with ethylene glycol, and on the morning of 27 January 2013 she tried to do just that.

She brewed him a special cup of Colombian coffee and laced it with ethylene glycol. George drank it, although at the time he thought it tasted rather sweet and afterwards he felt slightly drunk and his colleagues noted his slurred speech and unsteady gait. That evening he went to a restaurant with them and they noted that he could not hold his knife and fork, and one of them jokingly asked if he'd taken OxyContin, which is an opium-like painkiller that medical staff are known to take. He denied this, but he had enough wit to drive himself to the local hospital where he passed out. George was seriously ill and the cause soon came to light: ethylene glycol. This had damaged his kidneys, but he managed to survive. Ana was arrested, brought to trial, found guilty and sentenced to 10 years in jail, and fined $10 000.

1.6 ATTEMPTED MURDER, CRYSTAL PALACE, LONDON, 2013

Another would-be poisoner who was unaware of the antidote effects of alcohol was Jacqueline Patrick. She added it to her husband's favourite tipple, which was Cherry Lambrini. He was Douglas Patrick, a 72-year-old retired London bus driver. The drink is a sparkling cherry fruit wine flavoured with sugar and sweetener. It costs around £2.50 a bottle and it has an alcohol content of 5%. It was a drink with which the Patrick family decided to celebrate Christmas in 2013, and his wife Jacqueline (55) and their daughter Katherine (21) had bought a bottle just for him. They said they preferred other drinks with which to celebrate Christmas at their home in Crystal Palace, south London.

Wife and daughter had decided that their lives would be better if Patrick was out of the way. The previous autumn they had tried to poison him with ethylene glycol but all they had done was make him ill. Clearly they had not used enough to do the job. When it came to putting it in the Lambrini, they upped the dose considerably. Then they celebrated Christmas day by having a family row and Patrick drowned his sorrows in more drink. The amount they had added to the Lambrini that Christmas was enough to make him feel very drunk and he went to bed early.

On Boxing Day, Patrick was so ill that he needed medical assistance and he was admitted to King's College Hospital where he went into a coma, in which state he remained until 8 January 2014. On the way to the hospital, Mrs Patrick handed a typed note to the paramedics which said:

I Douglas Patrick do not wish to be revived as I would like to die with dignaty (*sic*) with my family by my side. D Patrick.

By the time Mr Patrick recovered, the doctors knew that he had been drinking ethylene glycol, although he denied any knowledge of this. He even told them he'd had a similar incident the previous October. It was enough for the police to be called, and the following day the two women were arrested. A bottle of antifreeze was found at the house. Also, incriminating text messages were found on their mobile phones. Messages dated

26 October, 17 November, and Christmas Day revealed their earlier attempts to kill him:

> I got the stuff and will give him some later – delete txt.
> He feels sick again I gave him more – delete this.
> Delete this but antifreeze is working.

Thankfully, Patrick did not die although he was seriously ill for a long time. Mrs Patrick was jailed for 15 years for each of the attempted murders while Katherine was jailed for three years on a charge of assisting her mother.

CHAPTER 2

Oxalic Acid and Murders in Manila and Liverpool

A word in **bold** *indicates that further information can be found in the Glossary. Only the first time the word appears in a chapter will it be so indicated.*

Previous generations were much afflicted by constipation, due in part to their exposure to lead and to a lack of fibre in their diet. Various laxatives were available, some of which are still used today, such as senna, but a popular remedy was rhubarb (Figure 2.1) and that worked because it contained the toxin **oxalic acid**, which irritated the gut into rejecting its contents. Rhubarb originated in China, where its benefits had been appreciated for 4000 years, and it first appeared in the UK in the 1770s.

Mrs Beeton's famous book *Household Management*, published in 1860, contained several recipes for pies and jams made with rhubarb, and she said that it was available in most kitchen gardens, as indeed it was. A popular dish was stewed rhubarb, sweetened with sugar and served with custard. Rhubarb contains several organic acids, of which malic acid predominates, but this is not toxic. Next comes oxalic acid, which can account for as much as 1% of the plant, with most of this being in the leaves. This is toxic.

More Molecules of Murder
By John Emsley

Published by the Royal Society of Chemistry, www.rsc.org

Figure 2.1 Fresh rhubarb contains the toxin oxalic acid. © Anna Shepulova/Shutterstock.

Although oxalic acid is available commercially, and dangerous if consumed, it is not one that a would-be murderer would choose to poison his or her victim with for reasons of taste. It would be almost impossible to feed it to someone without their knowledge such is its acidic nature, which causes it to attack the lining of the mouth and the gut. However, there are some examples of its misuse, as we shall see.

Oxalic acid has a habit of teaming up with calcium to form tiny insoluble crystals of calcium oxalate, which can be seen in the urine or which can grow to form painful kidney stones. The central nervous system is also affected by these chemicals. The body can get rid of oxalic acid so long as the intake of fluids is adequate, and for most people this is so. A large intake of oxalic acid, however, can prove fatal, as was found when children were given rhubarb leaves as a vegetable in World War I when food was in short supply.

2.1 OXALIC ACID

The Dutch botanist Herman Boerhaave extracted a salt of oxalic acid from sorrel. The acid itself was first produced by the Swedish chemist Scheele in 1776, and he made it by the chemical reaction of sugar and nitric acid. This is how it is manufactured today. An alternative method is to make its sodium salt by reacting cellulose with sodium hydroxide.

Oxalic acid comes as the dihydrate, with two molecules of water per oxalic acid molecule.† It is very soluble in water, to the extent of 150 grams dissolving in a litre. The solution so obtained is quite corrosive and used industrially to clean metals. It is particularly good at removing rust stains. It is also used in dyeing, where it acts as a mordant.‡ It is also used in tanning and for purifying oils and fats. Oxalic acid also has applications as a wood restorer because it will dissolve the top layer of wood to expose a clean fresh surface.

Oxalic acid occurs naturally in several plants that we eat. It is highest in Swiss chard (700 mg per 100 g), spinach (600 mg), rhubarb (500 mg), and beetroot (300 mg). There are also measurable amounts in tea, cocoa, chocolate, nuts, beans, and berries; peanuts, for example, have 150 mg per 100 g. A cup of tea will contain around 50 mg. Even though spinach contains a lot of iron, its high level of oxalic acid means that this mineral prefers to bind to oxalic acid and so is not easily absorbed by the body from this vegetable. The average person probably consumes 100–150 mg of oxalic acid per day. Patients undergoing dialysis have been found to release oxalic acid at higher-than-expected levels.

Plant cells can make use of oxalic acid, and it was always assumed that it had no role in animal cells and yet even human cells, such as blood cells, contain it in surprisingly large amounts. This seems to imply it has a role to play, although what this is is not yet known.

Oxalic acid became quite controversial in the USA earlier this century when it was claimed that the amounts in certain vegetables posed a health hazard. Since it has the ability to bind minerals like calcium, magnesium, and iron, it could result in a deficiency of these essential elements, which need to be absorbable from our diet. It was also associated with kidney and bladder stones, which can be very painful, and these may contain up to 80 per cent of calcium oxalate. They may also be the trigger of cancers in these organs. However, the body cannot completely avoid this acid because it is formed from any vitamin C that is

†The formula for this is $HO_2CCO_2H \cdot 2H_2O$ – see Glossary.

‡A mordant attaches itself to the material to be dyed and the dye molecule then attaches itself to the mordant.

surplus to requirements. The body cannot store this particular vitamin.

2.2 THE TOXICITY OF OXALIC ACID

The symptoms of acute oxalic acid poisoning are burning in the mouth and throat, abdominal pain, nausea, vomiting, and diarrhoea. Exposure to less-than-toxic amounts is said to cause a general weakness and affect the heart, and in the long term, it could lead to heart attack, convulsions, and coma. It is thought to exacerbate conditions such as gout and arthritis.

The **LD_{50}** of oxalic acid for humans is estimated to be 375 mg per kg, which means that a dose of around 25 grams would prove fatal to most people, but this would be equivalent to a dessert spoonful and would taste even more bitter than **Epsom salts** (magnesium sulfate). However, accidents have happened, as in Scotland at a mental institution in 1956 when a nurse was instructed to give two patients a dose of Epsom salts to treat their constipation. Mistakenly, she gave them oxalic acid instead – which looks very like Epsom salts – and they both died.

Oxalic acid causes death due to its combining with the vital metal calcium, depositing this as insoluble calcium oxalate (Figure 2.2) and thereby interfering with the key role of calcium in muscle action. It affects all muscle cells in the body, including those of the heart. The person affected will also experience a

Figure 2.2 Calcium oxalate crystals. © toeytoey/Shutterstock.

feeling of being paralysed, with tingling in hands and legs. For a while, the body will strive to replace the lost calcium, although this might just prolong the inevitable outcome.

The main symptoms of oxalic acid poisoning are swelling of the stomach, intense thirst, and vomiting dark brown fluid which looks rather like coffee grounds. Someone who has taken or been given a large dose of oxalic acid may live for up to two days, but most will generally die within a few hours. Those given a lesser dose may linger for up to two weeks.

2.3 ATTEMPTED MURDER, LIVERPOOL, 1852

The scene is set in 1852 in Hartley Court, King Street, which was in the area of Liverpool known as Kirkdale. The main characters were two unmarried sisters, Sarah and Ann Rimmer, and Sarah's daughter, 19-year-old Elizabeth, who was proving difficult. She had recently lost her job as a servant and hadn't found a new position, nor did she appear to be looking very hard for one, and this was beginning to annoy her mother and her aunt. At the end of March, she was sent out once again to find a job but apparently without success. When she returned home that afternoon her mother was furious and yelled "If you don't leave home I'll give you poison, or try some other means" or so said a neighbour, William Blundell, who lived in the same court and who overheard them arguing.

The following day, Wednesday 31 March, when Elizabeth came down for breakfast – late as usual – she found her mother had made a pot of tea, which was being kept warm in the oven. She drank a cup of it and, although it tasted very sweet, it burned her throat and stomach. She challenged her mother about it but her mother then took a sip of the tea, declaring that some **alum** must have somehow got into it.[§]

Elizabeth suspected she had been poisoned so she took some of the tea round to William Blundell and told him of her suspicions. He took her to a pharmacy in Castle Street where the pharmacist, Mr Haywood, tasted it and deduced that it contained oxalic acid. He gave Elizabeth some magnesium carbonate as an antidote and told her to go round to a Dr Arnold whose

---

[§]In those days, alum had various uses in the home; it would stop bleeding, it was a component of baking powder, and it had a toning effect on the skin.

practice was nearby and tell him what had happened. The doctor gave her a bottle of medicine to relieve the pain of her throat and stomach. Meanwhile, Blundell informed the police of what had happened.

When Elizabeth got home, there were two police officers in the house. Her aunt, Ann, then snatched the medicine from Elizabeth and hit her across the face, saying "I haven't done with you my girl. You ought to be hung, but hanging would be too good for you." One of the police was a Detective Murphy who searched the house and found oxalic acid in the fireplace of an upstairs room. He challenged Ann about it and she admitted it was hers but said "If they give me 12 months I won't care." Both she and her sister were arrested.

Elizabeth now made an even more serious charge against her mother and aunt. She claimed that her mother had given birth to an illegitimate child in July 1849, which may have been stillborn but more likely had been smothered by Sarah and Ann and buried in the garden of the house in Ormskirk where they were living. When the police investigated the garden, she pointed out the place where the baby had been buried. They found what appeared to be a dark sticky mass at a depth of three feet but there were no bones, although a surgeon whom they consulted said that the bones in a newly-born baby would quickly decompose.

Elizabeth also said they may have poisoned her grandmother in October 1851. She died suddenly and was hastily buried in the local churchyard.

The Rimmer sisters faced trial on 20 August 1852 at Liverpool Crown Court. When Elizabeth gave evidence she said that they had frequently used foul language to her, and a particular occasion was on the 31 March when a quarrel began about the tea not being ready when her mother and aunt came home from work. She also said that when she couldn't get a job they suggested that she should become a prostitute and earn some money that way, but this she refused to do. When giving this part of her evidence, Elizabeth was so overcome with emotion that she passed out and a doctor was called to revive her.

Other evidence was given of Sarah disposing of the contents of the teapot in the gutter outside the flat where they lived. Although no one had died, the two women now faced the death penalty, as the judge explained when he came to sum up the case for the jury. A recent Act of Parliament said that whoever administered poison with the intent to murder was to be treated as guilty of the offence of murder, irrespective of whether the intended victim had died or not. The jury found the two sisters guilty and they were duly sentenced to death. It was said that they seemed indifferent to their fate. In any event, the sentences were commuted to life imprisonment.

2.4 MURDER, MANILA, 2015

In the Sampaloc district of Manila in the Philippines was the Ergo Cha Tea Shop run by William Abrigo. It was there that Suzaine Dagohoy and her partner Arnold Aydalla sometimes came for a drink and to relax. On the morning of 9 April 2015 they ordered Hokkaido tea, a kind of tea that is very creamy and sweet and much favoured in Japan. It is prepared from Hokkaido milk, which is famed for its creaminess, and this is boiled with black tea leaves along with either sugar or honey. It is then left to cool before being served. The drink is intensely sweet and is named after Hokkaido, the northernmost island of Japan, which is famous for its natural hot springs, skiing, and its Daisetsuzuan National Park. The milk from cows which grazed in this part of Japan is reputed to have special health benefits, seemingly because the cows lead tranquil lives. The milk is also used to make chocolate, desserts, and a special kind of bread.

The owner, Abrigo, had made the Hokkaido tea himself and in the usual way, but when Aydalla and Dagohoy drank it that morning they found it rather unpleasant. They called Abrigo over and he also drank some of it and said he found nothing wrong with it. He also took a fresh cup of the tea and drank that as well. Within minutes, Dagohoy had started to vomit and then collapsed and was rushed to hospital, followed soon after by her partner Arnold. Abrigo, the teashop owner, also collapsed and he too was rushed to hospital. Of the three victims, only Aydalla survived; Dagohoy and Abrigo died later that day.

So who had put oxalic acid in the tea? The helper in the shop said that Abrigo's son Lloyd, who was studying accountancy at university, had brought some suspicious-looking fluid into the shop the previous day and that it smelled of bleach. The shop's CCTV showed that he had been wearing surgical gloves at the time, and he made a point of cleaning the things associated with the Hokkaido tea. Somewhat oddly, although Lloyd learned of his father's death later that day, he did not inform his mother of what had happened for several hours.

At the hospital, the doctors treating the three affected people at first thought they had been affected by the milk, which might have been contaminated. It soon became clear that something more deadly was responsible for their condition, because two of them died soon after they were admitted to the hospital. Clearly they had been poisoned. Tests were carried out for arsenic and cyanide but these proved negative. The Philippine National Police (NLP) were called to investigate and on 11 May they announced that oxalic acid had been found in the blood and stomach of the two victims. Another team of analytical chemists tested other things from the tea shop for oxalic acid and found it in the tea and the syrup used to make the Hokkaido tea.

The 23-year-old son, Lloyd, was the chief suspect and it was noted that he had taken particular care to clean up the kitchen area before police investigations could be made. He denied the allegations but on 15 May he was charged with the murder of his father and Dagohoy and the attempted murder of Aydalla. However, he was never to be tried.

In December 2015, it was decided by the city prosecutor that Lloyd was unlikely to be the perpetrator because he had no motive for killing the couple and, in any case, he did not prepare the milk tea. Suspicion also fell on the assistant in the shop whom Abrigo had been on the point of dismissing because he was dissatisfied with his performance, or so Abrigo's family said. In the end, the legal authorities decided not to proceed against Lloyd, and the case was dismissed and he walked free. So who did add oxalic acid to the tea that day? We may never know. The tea shop itself was closed down.

CHAPTER 3

Acrylamide in Fried Foods and in Auckland

A word in **bold** *indicates that further information can be found in the Glossary. Only the first time the word appears in a chapter will it be so indicated.*

Many of us start the day with either a bowl of cereal or a slice of toast, and as such, we consume a little **acrylamide**, which is a deadly poison. Later in the day we may eat some chips or fries, again taking in more acrylamide. Of course, the amount of this poison in such foods is incredibly tiny and not enough to do harm, although there have been studies suggesting this chemical might cause cancer. Acrylamide forms when a carbohydrate-rich food is heated to a high temperature, as happens when we roast, toast, grill, or fry something. However, the amounts of acrylamide that we consume are only measured in millionths of a gram and are easily disposed of by the body.

Could this chemical be used to poison someone deliberately and with impunity? For that to happen it would require several milligrams, and maybe even gram quantities, to enter the body, and then it would be obvious from the amount that is present that they have been poisoned. However, this is not easy to prove because acrylamide is speedily metabolised, and within a day it

More Molecules of Murder
By John Emsley

Published by the Royal Society of Chemistry, www.rsc.org

has disappeared. But acrylamide has an advantage as a poisonous agent because it can penetrate the skin so it need not just be delivered *via* drink or food.

It sounds like a perfect poison with which to attack someone, provided you can get hold of it, and that is not easy unless you work in a laboratory. There has been only one famous case in which it featured and even then it failed to kill, although it left its victim permanently disabled. But who would do such a thing? How would they do it? And what would be the reason?

3.1 ACRYLAMIDE

Around two million tonnes of acrylamide are made every year, with 45% of this being produced in China, followed by the USA with 20%, and Europe with 15%. It is a most versatile chemical in its polymer form, being used for things as diverse as grouting for tunnels and contact lenses. Most polyacrylamide is used in water treatment where it acts as a flocculator; in other words, it attracts suspended solids and allows these to be removed and so leave the water crystal clear. Another major use is in drilling for oil when a small amount of polyacrylamide can make water very viscous and when this is injected down a well it increases the output of oil. Paper mills also use it in the sizing of paper and the manufacture of cardboard.

On a smaller scale, polyacrylamide is used by famers and horticulturalists to prevent soil erosion, and it can be added to pesticides so that when these are sprayed on crops no spray drifts on to neighbouring land.

Acrylamide grout was introduced in the 1950s and was ideal for preventing water leaking into underground workings. The grout is applied as a sealant and is easy to work with and forms a stiff but flexible waterproof gel. It has been used in tunnel construction, and it was this use in Sweden which brought its dangers to the attention of scientists. One consignment of grout contained a significant amount of unreacted acrylamide monomer which then affected some of those who worked with it. Research showed that the extent of exposure was all important. When monkeys were dosed with acrylamide it was found that being regularly given 10 mg per kg body weight per day caused weakness in their legs and arms. However, when they were given daily doses of 3 mg per kg per day for a year they exhibited no signs of poisoning at all.

In 1967 a group of six workers were taken ill in the UK with unusual symptoms and the cause was due to their having handled acrylamide. Their hands were excessively moist and the skin was peeling off. In fact, the more their hands sweated, the more acrylamide was absorbed from the grout, and some of them handled it with bare hands, being totally unaware of the threat it posed to their health. The men also suffered other symptoms, such as being unsteady on their feet and confused in their speech.

Exposure to a lot of acrylamide leads to trembling, mental confusion, and difficulty speaking, and as it interferes with the nervous system, there will eventually be convulsions and loss of weight. When the men were removed from contact with the chemical it took up to 12 months before some of them were completely free of symptoms.

Other instances of workers being affected came to light in the 1990s, and it came to the attention of the medical authorities. Around 200 workers in Sweden who used this grouting material had been affected by it, and 23 were particularly badly affected. They exhibited symptoms that indicated their central nervous system was being targeted, beginning with numbness of the hands. The same thing happened in Norway during the construction of a tunnel to the new Oslo airport. The cause was eventually traced to about 5% of unreacted acrylamide in the material they were working with.

3.2 ACRYLAMIDE: ITS TOXICITY AND ITS DETECTION IN THE HUMAN BODY

Acrylamide poisoning causes a person to lose control of their bodily movements, and this is due to the effect of the molecule on parts of the brain, such as the cerebellum, and on the central nervous system. Victims will experience a feeling of pins and needles, sweating, muscle weakness, and weariness. Exposure to enough acrylamide to cause such symptoms is usually limited to those working with it in factory environments and on construction projects.

Once it enters the body, acrylamide mainly reacts with the amino acids **cysteine**, lysine and histidine. However, it is the reaction with cysteine that accounts for most of it. Acrylamide also poses a threat to DNA. The liver removes acrylamide by oxidising it to glycidamide with the help of the catalyst cytochrome P459 2E1. This then goes on to react with glutathione and haemoglobin, and the product of this latter reaction will remain within the body for up to four months.

The detection of acrylamide in the human body must involve measuring the products formed from these natural chemicals with which it has reacted. If it has attached itself to cysteine it forms CEC (carboxyethyl cysteine), which is excreted *via* the urine. However, this would have to be tested for soon after poisoning was suspected because all trace of acrylamide may be gone within two days. Another way of measuring the amount of acrylamide is by means of the haemoglobin molecules to which it has attached itself, and the adduct is referred to as CbEV-globin (CbEV is short for carbamoylethyl valine) or simply CEV. This adduct hangs around for longer. Even longer lasting are the traces of acrylamide in hair, which become a long-term record of its presence in the body (Figure 3.1).

Finding exactly where it is present along a strand of hair will reveal to within a day or two when the poison was absorbed, as we will discover below.†

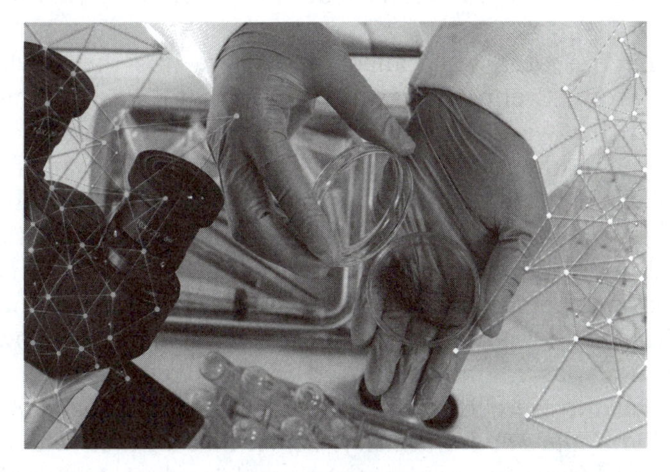

Figure 3.1 Analysis of hair such as would be used to detect acrylamide traces. © Tonhom1009/Shutterstock.

†Hair grows at a rate of 12.5 mm per month.

3.3 ACRYLAMIDE IN FOOD

Although the acrylamide monomer is a rather nasty chemical, it is also one which we produce in our own home when cooking. It forms during the preparation of foods that are baked and fried and, in particular, from potatoes and cereals. However, amounts are so tiny that this was not realised until the turn of the century. It all began in 1997 when famers living in the Bjare peninsula in southwest Sweden reported that something was affecting their livestock; cows were becoming paralysed and even dying, and dead fish were found floating in breeding ponds.

Margareta Törnqvist of the Department of Environmental Chemistry at Stockholm University was called upon to investigate, and she found that the cause was a high level of acrylamide monomer in the sealant used in a tunnel being constructed and from which it had leaked into ground water. It also affected the men working in the tunnel, some of whom were clearly ill. Their exposure to the poison was assessed by measuring the amount of CEV in their blood. To check her method of analysis, and for comparison reasons, she also took blood samples from a group of people who had not been affected, to act as a reference group. In fact, she had expected there to be no CEV in their blood but there it was: they too must have been exposed to acrylamide. But how?

Further research was called for. Where was the acrylamide coming from? Eventually, it was realised that it was coming from the food they were eating. Tests on rats showed that those fed fried food had much higher levels of CEV than rats fed the uncooked food. At the time, her group's results were published in the journal *Chemical Research in Toxicology*, but she was persuaded not to publicise her findings *via* the media until further research had been carried out. More research was clearly needed, and indeed, her group were subsequently able to show that acrylamide was mainly produced in fried potato-based foods. These results were then published in 2002 in the *Journal of Agricultural and Food Chemistry*, which had a much wider readership, and they came to the attention of Sweden's National Food Administration.

They began their own tests and confirmed that all kinds of cooked and fried foods contained small amounts of acrylamide.

These included crisps, chips, fries, biscuits, crackers, and breakfast cereals.[‡] The amounts varied even for the same kind of product, so for crisps it could range from 330 to 2300 micrograms per kg, and the amount was influenced most by the cooking temperature. Once this was realised, then people began to take action, and when similar foods were analysed in New Zealand in 2012, it was found that crisps had a level of 580 micrograms per kg compared to a level three times as high in 2006. Wheat-based breakfast cereals had 270 micrograms per kg. The overall dietary intake of acrylamide in 2012 was assessed as being between 0.7 and 1.0 microgram per kilogram body weight per day for adults.

More worrying was the evidence that acrylamide could cause cancer. The Swedish scientists also carried out tests on laboratory animals and showed this to be so; the implication being that human cancers of the gut might also be caused by it. Because so many people have been eating small amounts of acrylamide for most of their life it appeared as though the risk was small, and there was no possible way of knowing whether a person with cancer had been exposed to above-average amounts of this chemical so it was impossible to link it definitely to human cancers. It was also noted that men exposed to acrylamide at work were no more likely to suffer from cancer than the population at large. But this finding was only announced to counter a media scare promoted by environmentalists which was causing worldwide alarm.

Soon it was shown that acrylamide was present in more than just fried foods; it was also in black olives, coffee, and dried fruits. However, the World Health Organization (WHO) concluded that for this to present a serious health risk, intake of this chemical would have to be 500 times higher than the average intake of a normal person. Even so, and under threat of a lawsuit, several major US food manufacturers agreed to reduce the level of acrylamide in their products. Meanwhile, sceptics pointed out that a person would have to eat their own body weight of chips per day to get even one tenth of the amount considered to put their health at risk. Even so, it prompted

[‡]There may be some confusion here because these words have different meanings. I take crisps to mean the thinly sliced fried potato snack, chips to mean potatoes that have been cut into finger-sized pieces and fried, and fries to be those that are thinner and which are made from extruded mashed potato which is then fried.

research into ways in which the amount of acrylamide in food preparation could be reduced: toast should be only lightly-toasted and chips should be only a light golden colour and not dark brown.

3.4 ATTEMPTED MURDER, AUCKLAND, 1992

Those who are elected to be Fellows of the Royal Society of London can be proud of their achievement. One such person was New Zealand's Professor David Lloyd, who became an FRS in 1992. Later that same year he was poisoned with acrylamide. He survived, albeit badly damaged, and his case is the only one of suspected attempted murder with this noxious chemical that we shall look at in this chapter.

Lloyd was left a physical wreck but survived and lived another 14 years, finally to die 'peacefully' at his home in Christchurch on 30 May 2006. 'Peacefully' is what it says on the website of the Royal Society of New Zealand, which reports his lifetime of research but not his personal life. Lloyd had worked on the reproductive biology of plants and was able to show some plants of the same species have different genders, some have both genders in the same plant, and some even change gender from year to year. His ground-breaking research made him famous among biologists and it was on the basis of this that he was admitted to the Royal Society. However, to many he is better known because of his poisoning by acrylamide.

The story begins in 1992 when Lloyd was 55 years old and was the Professor of Plant Biology at the University of Canterbury on the south island of New Zealand. He had been married and had a family, but he had separated from his wife and now lived with a new partner, Vicky Calder. She was 10 years younger than him. She too had children by a previous relationship and they came to live with their mother at Lloyd's home in Cashmere, which is an up-market suburb of Canterbury.[§] Calder was a molecular biologist at the Christchurch School of Medicine.

Lloyd and Calder had been living together for six years when, in early 1992, Lloyd was told that he had been elected a Fellow of

[§]It was there that the famous whodunit novelist Ngaio Marsh had lived. She was born in 1895, made a Dame of the British Empire in 1966, and died in 1982. She was almost as famous and prolific as Agatha Christie.

the Royal Society and in June that year they flew to London for him to receive the award. They then went on a 10 day holiday touring Italy. At the end of that visit, Lloyd flew to the University of California in Berkeley to attend a conference while Calder went to attend one in Amsterdam.

While he was at Berkeley, Lloyd contacted Canadian-born Linda Newstrom whom he had previously known when he visited there back in November 1990. Then, she was a research associate studying botany, and he was engaged in a year-long research project at Cornell University, New York State. He had gone to Berkeley as a visiting professor and there he had met Newstrom who was mapping the distribution of trees of the tropical rain forest in Costa Rica. She had devised a system for classifying plants according to the way they flowered, and this was of particular interest to Lloyd.

The two of them spent quite a lot of time together and he made advances to her, but these were rebuffed because she knew that he already had a partner in New Zealand. In any case, she said that if they were to start a relationship it could only lead to gossip and maybe even put her career at risk. Nevertheless, they had fallen in love with each other; a love that was finally to bring them together in 1992.

In January 1991, Lloyd telephoned Newstrom to say he would be visiting California the following year and would like to see her. She was non-committal but she was interested in his work and they discussed that over the phone. Eighteen months later, Newstrom was much more receptive to Lloyd's advances when he arrived in California and they renewed their affair. He had found the love of his life and he wanted to finish with Calder. On 28 June, during the flight back to New Zealand, he decided to make a clean break.

Calder was at the airport waiting to greet him and sensed he was no longer the same man. When they reached home, he told her that he had found a new love and planned to spend the rest of his life with her. Calder could not believe what had happened, but it meant the end of their relationship. Lloyd moved out of their home and found himself a flat. He then rang Newstrom to tell her that he had left Calder.

Understandably, Calder was upset and wrote several letters to Lloyd accusing him of betraying her. She then began a campaign of revenge against her former partner; she stuck abusive notes on the windscreen of his car; she let down the tyres; she cut up all the shirts and trousers he had left at their home, although she did spare his 1964 Harvard PhD graduation gown.

She sent photos of herself and Lloyd to Newstrom in Berkeley, along with a letter saying "I knew that he did not love me. . .but I loved him and always will. Take David by all means but be careful. Don't be blinded by your love for him as I was." She also e-mailed academic friends around the world to tell them what had happened, and she made life difficult for Lloyd by stopping the sale of his house. He gave her six months to move out.

More puzzling was a letter that Newstrom received accusing her of breaking up Lloyd and Calder's relationship and saying her professional status was at risk because of it. The letter came from someone who signed herself as "Salia" but she was never traced and it seemed most likely that Calder had written it. She also sent a love letter to Lloyd from someone who signed herself "Sara Mills."

It's what happened in mid-December 1992 that is unclear and Lloyd was unable to remember when questioned later. No doubt Calder had discovered that Newstrom was due to fly out to Christchurch to spend Christmas with Lloyd. It appears he was looking forward to her visit as he told some friends when he entertained them to dinner at his flat on Friday 11 December. His daughter, who visited on the Saturday, found him fit and well but things were to change.

On Sunday 13 December, Lloyd became ill with excessive vomiting, which he told his daughter was probably due to food poisoning and he blamed some reheated leftovers that he had eaten. He told Newstrom about it when they spoke on the phone that evening. The next day he told her that he was still vomiting and feeling bad.

The following Thursday, 17 December, Lloyd rang the police to report a burglar or prowler outside his flat. PC Jeannie McCormick went round to investigate and found him in a distressed state and unable to walk. There was no sign of an intruder. She suggested he contact a doctor but he did not take her advice. Next day, a technician who worked with him at the university, Gerald Cuthbert, called to deliver his university mail and found him unable to stand and slurring his words. He immediately took him in his car to see his doctor who was already treating Lloyd for an enzyme deficiency condition, the enzyme being alpha-1-antitrypsin.¶ Lloyd was unsteady on his feet, talkative, delirious, and appeared to be hallucinating, but his symptoms were clearly not linked to his genetic disorder and the doctor immediately sent him to hospital.

Newstrom arrived as planned, hoping to spend Christmas in Christchurch, and she stayed in Lloyd's flat. She went to see him in hospital and there she met Calder, who was also visiting him. The two women agreed to visit Lloyd on alternate days. Then something happened that upset Newstrom. Among Lloyd's correspondence was another letter from Sarah Mills, who was clearly one of his lovers. Who was she? The letter was sent back to the Australian address from which it appeared to have been written along with a note saying that Lloyd was very ill. The letter and note eventually came back undelivered, "address unknown."

On 4 January, another postcard from Sarah Mills arrived at Lloyd's flat, and it convinced Linda that her lover had double-crossed her. So she searched for further evidence of this secret love affair but found nothing. When she checked with his phone company, she discovered that Lloyd had never made any phone calls to Australia. She suspected Vicky Calder was responsible and she told the police. Detectives went to Christchurch Medical School to interview Calder and, while she admitted to putting notes on his car's windscreen and to cutting up his clothes, she claimed that she had not seen Lloyd until she heard of his illness and went to visit him in hospital. There things rested for the time being but investigations were being made by the detectives, and they became convinced that a crime had been committed against Lloyd and that the person responsible was Calder.

In theory, Lloyd's illness could have been a natural condition but it seemed more than likely that it had been caused by a poison. In January 1994, new testing methods supported the latter theory and indicated acrylamide as the toxic agent. To confirm their suspicions, they sent samples to England for further forensic analysis by toxicologist Dr Pamela Le Quesne. She was a consultant neurologist based in Middlesex, UK, and

¶Lack of this affects the lungs and can cause cirrhosis of the liver.

she confirmed that Lloyd's early symptoms were consistent with acrylamide poisoning. By 6 June the authorities had decided that only one person could have done this and that was Calder, and she was arrested and charged with attempted murder.

Calder's trial began at the Christchurch High Court on Monday 18 September 1995 before a jury composed of 10 men and two women. From the start of the trial, the prosecution admitted that no one had actually seen Calder and Lloyd together on that fateful weekend back in December 1992, but he said that the case against her would be proved by circumstantial evidence. The first witness he called was Linda Newstrom, who told of her meeting with Lloyd, falling in love with him, and how she was devastated by what had happened to him. It was also clear that she had been deeply upset by the Sarah Mills letters.

Lloyd did not appear at the trial but provided a short statement that he was not able to remember what happened that fateful week in December 1992 or about his relationship with Calder.

Then followed a series of witnesses who had been colleagues and associates of Calder, and they recounted some of the conversations they had had with her. One said that Calder had told her that she would get her revenge on Lloyd by using a poison which would not be discovered, although the witness could not remember what that was. Another witness, a technical officer, recounted how Calder said she had a poison which she would smear on door handles and which could be absorbed through the skin and was untraceable. The witness said that she did not take these comments seriously but when Lloyd became ill she even feared for her own safety and began wiping her own door handles.

Another witness, a botanist, recounted Calder talking about untraceable poisons at a dinner party. A medical student who had worked with Calder also told of her interest in untraceable poisons, although he could not remember those she had listed except that one was a common laboratory chemical. The witness also recalled a party he had attended on Saturday 12 December at which Calder was present and that during it she had gone out for some time, ostensibly to buy more drink. All this kind of evidence was questioned by the defence as unreliable because it relied on personal memories.

The court was told that there had been three phone calls between Lloyd and Calder on that December day, showing that they had been in contact, but these calls were relatively short in duration. Then there was the fake letter from Sarah Mills. An Australian woman of this name had been found but she denied any knowledge of Lloyd. A postal official confirmed that the stamps and franking marks on the envelope were suspicious.

Forensic pathologist Martin Sage of Canterbury University had been the first person that doctors turned to when they began to suspect Lloyd had been poisoned. Although Sage could not say for certain what was affecting Lloyd, he said it was likely that he had been poisoned with something. Samples of Lloyd's hair were sent for testing for heavy metals, as it was thought he might have been poisoned with one of those, such as mercury or thallium, but those tests proved negative. However, Sage eventually came up with a likely cause when he read something in the published literature and that put acrylamide into the frame.

Hair analysis was a key type of forensic evidence presented in the Calder case. The forensic science on this was performed by Eric Cairns, and he was able to show the presence of CEC and, moreover, he was able to compare the amount he found with the amount of CEC that is present in the hair of people who work with acrylamide. The level in Lloyd's hair was 100 times greater than that. Moreover, Cairns was able to show that the high level corresponded to the presence of acrylamide in Lloyd's body in mid-December, judging from its position in the hair. He was able to say that Lloyd had received a large dose of acrylamide around the middle of the month but his results showed the presence of this chemical earlier in the month, albeit at low levels.[‖]

Counsel for the defence now tried to show that Cairns' evidence fell short of what was required, and Cairns was kept in the witness box for six hours. The barrister knew of another test for acrylamide, detecting CEV, and this had recently been developed. Why hadn't he used that? In fact, the New Zealand Institute of Environmental Science and Research had tried the new test but come up with conflicting information. The prosecution had hoped to clarify matters by calling Professor

[‖]The presence of small amounts of acrylamide in other part of Lloyd's hair would have come from his diet, something that was not realised at the time of the trial in 1995.

Norman Aldridge, a renowned UK toxicologist, who'd looked at the analysis results for acrylamide in hair, but he became too ill to travel before the trial and indeed died a few months later. Instead, they called Dr Martin Johnson, also from the UK, and he revealed to the court that he'd tasted acrylamide crystals and said they had no taste. What he was able to tell the court was that it appeared that Lloyd had been given more than one dose of the poison. It would appear he had one dose on Saturday 12 December and then another dose the following week.

The prosecution had 60 witnesses take the stand during the first three weeks of the trial, many of whom were aggressively questioned by Calder's defence lawyers. Now it was their turn to offer alternative evidence, which they did. However, Calder was not called to the witness box, which always seems suspicious in itself, but the judge told the jury not to read anything into this. One of the main witnesses was also from the UK, Hugh Rushton, an expert in hair growth, and he refuted the prosecution's claims, saying that the forensic analysis had been rather poor and the jury should not rely on their results.

Now it was the turn of the defence and their strongest witness was also from the UK, Dr Moira Shelley of Manchester. She had treated a young man who had tried to commit suicide with acrylamide but had survived and, unlike Lloyd, he had recovered completely, to the extent that he was again involved in sport. Because Lloyd's symptoms appeared so different, the implication was these were not due to acrylamide poisoning. Another witness said that the reason Lloyd was ill in December was probably due to a stomach bug that triggered his immune system, but the prosecution was able to show that this witness had had no hands-on experience of this condition and knew nothing about acrylamide. However, those affected by acrylamide were so few in number that there could be no certainty about its symptoms in a particular individual.

In the summing up at the end of the trial, the prosecution claimed that only Calder had the motive to want to harm her former lover and that she was a "cocktail of emotions" when she poisoned him. There were love, hate, anger, bitterness, and revenge. The defence said that they had provided enough counter-evidence for a verdict of not guilty. For two days, the jury debated the evidence and then reported to the court that they could not agree a verdict, and they were discharged. There had to be a second trial and that took place in December 1995.

3.4.1 The Second Trial

At the second trial, evidence presented in court was based on the analysis of CEV in blood samples taken from Lloyd at the time he first became ill. CEV can persist for up to four months and this showed that Lloyd had 800 times more of this than a normal individual, and Lloyd had absorbed acrylamide over a period of two weeks in December 1992. However, when it comes to presenting forensic evidence in court, if this is regarded as experimental science rather than well-established science it will not be admitted. The judge decided that, for the CEV results, the method of analysis was acceptable and he approved of the evidence being given.

The Crown had also consulted an expert in Stockholm, Sweden, Dr Anti Kautiainen, who told the new jury that the level of CEV in Lloyd's blood was 860 times higher than those in the control group. The prosecution had also found a nurse at the Christchurch Hospital who was there when Calder visited Lloyd at 8 pm one evening and she had even requested to spend time alone with Lloyd. When the nurse returned, it was to find Lloyd in pain and Calder helping him to drink water from a glass. The detectives also found someone who lived near Lloyd's flat who remembered seeing a red Honda Civic parked outside in mid-December 1992, and that was the make and colour of Calder's car. Of course, this proved nothing, as the defence were quick to point out. In any case, the defence also had a new witness from the USA who claimed that Lloyd's symptoms did not fit with acrylamide poisoning.

The second jury deliberated for about nine hours and eventually came up with a verdict of not guilty, and that was that. The result was inevitable according to a review of the forensic evidence by Danielle Wornes of Murdoch University, Perth, Australia, published in the *Journal of Forensic Science & Criminology* in 2016. She says that there was no reliable technique for gathering the evidence of acrylamide poisoning and that this was bound to confuse a jury made up of ordinary members of the public. However, we cannot know whether the two juries were baffled by the forensic evidence or, if they did understand the science, they were just not convinced by it. In any case, it seems that Calder was not guilty of attempted murder and so ended a most remarkable case.

And what happened to the main characters of this drama? Lloyd never fully recovered his health, although on 18 June 1994 he had married Newstrom. He died on 20 July 2006 aged 68. Calder went on to become Vicky Webb and she joined the National Institute for Water and Atmospheric Research in Wellington in 1998 and led several science programmes, including one which studied diseases among farmed fish.

CHAPTER 4

Difenacoum, Amitriptyline and York

A word in **bold** *indicates that further information can be found in the Glossary. Only the first time the word appears in a chapter will it be so indicated.*

What will kill a rat may also kill a human, so it is perhaps not surprising that murderers over the years have resorted to rat poison to solve their problems. Such a product is relatively easy to obtain because it is needed if you are involved in a business where vermin like rats are a threat. One such product in the UK is Neosorexa, which can even be bought from Amazon, and its sale is possible because it contains very little of a powerful ingredient, **difenacoum.** This is fatal to rats and mice but not to humans, unless of course they deliberately consume a large amount. You would need to digest several hundred grams of Neosorexa to put your life in danger because it contains only 0.005% of the toxic agent, and even then there is a simple antidote for the poison and that is vitamin K.

When difenacoum is sold as a rat poison, it is mixed with sugar to make it attractive to eat and hydrocarbon wax (of the type used to make candles) to bulk it out. Then it is coloured blue.

In this chapter I have also included another chemical, **amitriptyline**, because of a famous poisoning case in which the

More Molecules of Murder
By John Emsley

Published by the Royal Society of Chemistry, www.rsc.org

would-be murderer decided to use this common antidepressant when her attempts to kill her husband with rat poison failed. Because toxic ingredients are now so strictly regulated, it is almost impossible to obtain any that will kill, but there are some medicaments that might kill if someone can be persuaded to take an overdose. Even that may not work, as we shall see when we discuss dear Heather Mook, a successful liar and thief, but not always a successful poisoner.

4.1 DIFENACOUM

Difenacoum was introduced in 1976 and is referred to as a second generation anticoagulant rodenticide. It is sold under trade names such as Ratak and Neosorexa. It is deadly because it prevents blood from clotting and the animal bleeds to death. It has been described as super-warfarin. Warfarin was the first-generation anticoagulant to be discovered and was introduced as a rat poison in 1948; it was very effective to begin with. Since then, some rats have developed resistance to its effect and it has been replaced by difenacoum, which is a much more powerful anticoagulant. Both work by blocking vitamin K. This ability to prevent blood clots was eventually to find use with humans, and warfarin is prescribed to those at risk of thrombosis, when a blood clot can block a vital artery in the brain, which may cause a stroke.

By means of radioactive isotope analysis, difenacoum has been shown to concentrate mainly in the liver and the pancreas, but it invades many organs of the body when it has been ingested. Small doses of difenacoum cause nose bleeds as well as bleeding gums and there is easy bruising. Large doses will cause bleeding in the stomach and from the rectum. Indeed, many internal parts of the body can be so affected. The symptoms of difenacoum may not appear for a few days after it has been ingested.

The **LD_{50}** of difenacoum for rats is 2 mg per kg, for dogs is 50 mg, and for cats is 100 mg. If the LD_{50} for humans is like that for rats then 140 mg might well prove fatal, although if it were like that for dogs it would require 7 grams to put life at risk.

Warfarin and difenacoum decrease blood coagulation by blocking an enzyme called vitamin K epoxide reductase. This enzyme recycles vitamin K after it has played its role in causing blood to coagulate to stop bleeding. Difenacoum is a more potent and persistent antagonist of vitamin K than warfarin. Those intending to commit suicide with it need to take large doses but if they are caught in time they can be given **phytonadione**[†] (vitamin K) and they will recover, although it may take several hours for this to take effect and thereby restore the clotting ability of blood. During this time the patient may need to be given a blood transfusion. Children who eat the rat poison can similarly be treated. Those who have been affected may need to be treated for up to three months to make a full recovery.

A 17-year-old girl tried to commit suicide in May 1981 by eating 500 grams of Neosorexa, which contained around 25 mg of difenacoum. (She also consumed broken razor blades and map pins.) She was treated with vitamin K and the normal clotting action of her blood was restored after 30 days. She survived.

4.2 AMITRIPTYLINE

Amitriptyline (Figure 4.1) is a commonly prescribed medication, mainly given to those with mental conditions such as depression and anxiety. It can also be prescribed to relieve irritable bowel syndrome and migraine. It is one of a group of medicines known as tricyclic antidepressants (TCAs), so named because their molecular structure consists of three interlinked rings of carbon

Figure 4.1 Amitriptyline, a commonly prescribed drug. © Sakonboon Sansri/Shutterstock.

[†]Known as phytomenadione in the UK.

atoms.[‡] Amitriptyline was discovered in 1960 and approved by the US Food and Drugs Administration (FDA) the following year, and it is on the World Health Organization (WHO) list of essential medications.

A typical dose of amitriptyline, as prescribed by a doctor, is 10 or 25 mg. The LD_{50} for amitriptyline is estimated to be about 6.5 mg per kg so that for a normal 70 kg person an intake of 0.45 grams (450 mg) could prove fatal. Lesser amounts could also put a person's life at risk.

Overdosing on this medication will lead to excessive dizziness and possibly to low blood pressure, an erratic heartbeat, seizures, and hallucinations, of which seizures and the effect on the heart are the most life-threatening. Amitriptyline works its wonders, or its damage, by inhibiting the reuptake of the neurotransmitters serotonin and noradrenaline (norepinephrine) from the synaptic cleft into the nerve. This prolongs their action at the synaptic cleft. There is no specific antidote to amitriptyline poisoning. Activated charcoal is the first line of defence, provided it is given soon after ingestion, with other treatments given as necessary, such as anticonvulsants like **diazepam** and the beta-blocker propranolol.

4.3 MURDER, HENAN PROVINCE, CHINA, 2007

This murder really was literally the kiss of death for the victim.

Xia Xinfeng and Mao Ansheng lived in the village of Maolou, which is in Henan Province, central China, and near Xinxiang City. For years they had been lovers, and indeed at the start of their relationship they had vowed eternal love and agreed that if either of them cheated on the other then he or she would have to die – or at least that's what Xia claimed when she was arrested for Mao's murder. Whether the lovers really did make such a pact is debatable.

On 8 January 2007, Xia observed Mao flirting with another woman and she suspected the worst because he seemed to have lost interest in his intimate relationship with herself. He deserved to die and she decided to kill him, and in a way unique in

[‡]These have been used to commit suicide and account for a few fatal poisonings every year.

the annals of crime: she would pass a fatal dose of difenacoum to her lover *via* a passionate kiss.

The following day she ground up some rat poison, which is sold in China in a form that has a high level of difenacoum, and put it in a tiny capsule which normally contained a painkiller. Then she took it with her and met up with Mao. She slipped the capsule under her tongue and greeted him with a long passionate kiss during which she worked the capsule into his mouth and remained kissing him until he had swallowed it. He quickly became ill and was admitted to hospital, and there he died a long and painful death a few days later.

Xia was found guilty by the court, sentenced to death, and hanged.

4.4 ATTEMPTED MURDER, YORK, 2007

Heather Templeton was born in 1950. She was an outgoing individual who made friends easily and charmed them with her conversation, although this was often a pack of lies. Those who believed her ended up losing thousands of pounds. Detective Inspector Nigel Costello led the police investigations into her behaviour in 2007, which resulted in her being put on trial for attempted murder. The detective later said that he was impressed by the meticulous way she planned and carried out her many frauds. However, it is Heather's career as a poisoner which is of interest here, although this activity was a relatively minor part of her criminal lifestyle.

In 1980, Heather had been put on trial for killing her 10-month-old daughter, but the baby's death was eventually attributed to cot death. This was at a time when cot death was a regular feature in the media, being blamed on chemicals used to make cot mattresses.[§]

In 1981, she tried to poison her seven-year-old daughter with amitriptyline. At the time, she was living in Manchester and was then called Heather Booth. One day she ground up tablets of amitriptyline and mixed them into the girl's food and drink, but her murder attempt failed and the girl had to go into hospital.

[§]A Government report eventually concluded that such claims were alarmist scaremongering, but not before millions of such mattresses had gone to landfill.

When the cause of her condition was revealed, Heather was arrested. She had also stolen money from her husband and was eventually taken to court and given a 12-month prison sentence, suspended for two years. She pleaded that what she had done was a cry for help and the court believed her, hence the lenient sentence. Her husband was less forgiving and divorced her.

In 1988, Heather was now Mrs Hope, and she and her new husband ran a wine bar in Stamford, Lincolnshire. There she stole from the catering manager whom she had asked to help her financially because she was temporarily short of cash. She persuaded him to give her a signed blank cheque saying she would fill the amount in later when she knew what she really needed, but she assured him that it would be for a relatively small amount. In fact, it amounted to £12 000. She was jailed for six months.

Then she moved to Peterborough where she told friends and neighbours that she was heavily involved in the Scotch whisky firm Whyte and Mackay, and persuaded them to lend her money. One wealthy friend gave her a blank cheque, and then found that she used it not to help fund her schemes but to buy several top-of-the-range BMW cars.

Another neighbour gave her £37 000 to buy a car for them, and this she delivered to their door a few days later. In fact, she had not bought the car but had hired it, something they only discovered later. Meanwhile, Heather pocketed the money and quickly spent it. She told other friends that she was planning to buy a hotel for £2.8 million and intended to spend £2 million renovating it, and would they like to become involved. Some believed her and so lost money. All in all, her swindles are thought to have amounted to £5 million. She eventually found herself in the dock of Peterborough Crown Court and was jailed for three years. Another divorce followed.

Next, she moved to the hamlet of Middlethorpe on the outskirts of York and which is famous for its imposing 17th century hall, now used as a hotel. Again, she was soon robbing her new friends and neighbours. She had told them she was undergoing treatment for cancer and living off an allowance from her former husband. She also led them to believe that she owned property in Florida and earned money from that. She eventually conned them out of £20 000 before she was found out. At York

Crown Court, she admitted nine frauds and was again jailed for three years.

Our interest in thrice-married career criminal Heather begins in 1997. It was in that year that she met John Mook, a 58-year-old retired bus-driver and widower, who had three grown-up children. He was charmed by her. They got married and lived in his home in Heslington Road in York.[¶]

Mr Mook's mother, Freda, lived in the seaside town of Scarborough, but she was old and frail and John decided that it would be best if she sold her home and she came to live with them, which she did. However, it was not long before she fell out with her daughter-in-law, who persuaded John that his mother needed the special care that could only be provided by a private residential care home. They found one for her to which she agreed to move. However, she did not have a large enough pension income to pay their fees, but could pay them if she drew on the £43 000 she had got from selling her house.

Mr Mook did not like dealing with the financial side of things and was happy for Heather to take control of his finances and savings, which now included his mother's money. He thought that had been deposited in his bank account, but it was now being used to fund an extravagant lifestyle and holidays. The care home's fees were not being paid and old Mrs Mook eventually owed more than £10 000. Heather was now at risk of being found out, and so she decided to poison her husband before he learned the truth. That was early in 2007.

When a letter from the bank arrived and the scam was about to be exposed, Heather went out and bought some Ratak rat poison, but she did not grind it up fine enough to disguise it when she added it to some spaghetti Bolognese. When Mr Mook was eating it, he suddenly crunched something in his mouth which left such a bitter taste that he spat it out.[‖]

Clearly, difenacoum was not going to achieve the end that Heather wanted, so she next resorted to the antidepressant amitriptyline. This made John ill and his heart was racing. He was totally confused and having hallucinations. His doctor sent

---

[¶]Ironically, this road joins up with Thief Lane.

[‖]Ratak appears to contain a bittering agent which would deter humans from eating it but not deter vermin.

him to hospital. When his daughter Tracy paid him a visit, she found him lying on a mattress on the floor telling her that he was fishing and driving a bus. Soon, the doctors discovered the real reason for his condition and informed the authorities.

Heather admitted to having given her husband large doses of antidepressant on at least two occasions, and smaller doses on a more regular basis. It appears he often suffered from tiredness but attributed this to a prostate condition from which he was suffering. She told him that the pills would help him and he believed her. The forensic examination of Mr Mook's hair showed that he had been poisoned several times over a period of a month. It seemed that she had persuaded him to take as many as nine pills, telling him they were muscle relaxants. When his wife visited him in hospital she even gave him another antidepressant pill.

John Mook survived, and Heather was arrested and found guilty at York Crown Court of attempted murder and given an indeterminate jail sentence of at least five years. Heather admitted to 19 offences of deception relating to stealing her husband's and mother-in-law's money. After she was found guilty of attempting to murder John Mook, the police looked again into the death of her ten-month-old baby. It now seemed to them that back then she had literally got away with murder.

So ended the curious career of a woman who had a charming way with people but was a terrible thief and an incompetent murderer.

CHAPTER 5

Temazepam and the Man with a Murderous Plan

A word in **bold** *indicates that further information can be found in the Glossary. Only the first time the word appears in a chapter will it be so indicated.*

Temazepam was first made in the 1960s, and it became a prescription drug in 1964 when it was realised that it was an ideal treatment for insomnia. Eventually, it was to become one of the most widely prescribed drugs, and because it could deliver a pleasant high, it became one of the drugs traded on the street. Even so, temazepam can be deadly if consumed to excess; a fact that some would-be murderers have exploited, although not always with the result they were seeking.

There are several words used to describe the temazepam used as a recreational drug, such as King Kong pills, jellies, mazzies, and mommy's big helper. In the UK, it is a Class C controlled substance as defined by the Misuse of Drugs Act, and pharmacies need to show they have special facilities for storing it. Manufacturers in the UK have replaced the gel-capsules, which contained the drug in a powder form, with solid tablets.

More Molecules of Murder
By John Emsley

Published by the Royal Society of Chemistry, www.rsc.org

5.1 TEMAZEPAM AND ITS MEDICAL USES

Temazepam is chemically similar to **diazepam**[†] (Figure 5.1) and they are part of a group of drugs known as benzodiazepines, which have the same mode of action in the body. In fact, when the liver begins to deal with diazepam, it first adds a hydroxyl group (OH), thereby converting it to temazepam. Clearly, if temazepam is given as a medicament then it will avoid this step and its effect will be felt more rapidly and more intensely than other benzodiazepines, as a 1995 study found. Most of the drug is absorbed from the gut and excreted in the urine with some also appearing in the faeces.

Temazepam is known simply by its technical name in the UK, but in the USA it is better known as Restoril. Its action in the body is to boost the effects of gamma-aminobutyric acid (aka GABA) by binding to the GABA receptor and thereby controlling the movement of chloride ions into nerve cells. Its most obvious effect is to sedate the person who takes it. And, like many drugs, it causes side effects in some people, such as staggering when they walk, giddiness, muscle relaxation, headache, lethargy, and slurred speech, although relatively few people

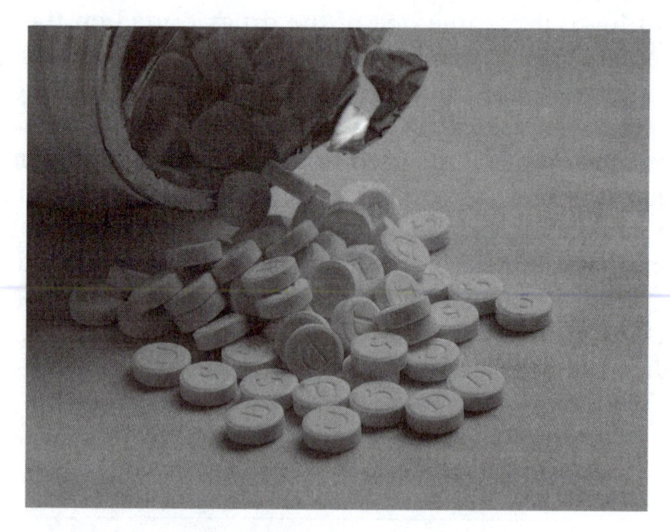

Figure 5.1 Diazepam pills. © Anukool Manoton/Shutterstock.

[†]Better known as Valium.

are affected this way when taking the prescribed dose of between 15 and 40 mg. Temazepam begins to be effective within 30 minutes of being taken.

Temazepam is prescribed for severe insomnia and treatment is generally limited to four weeks because of possible dependency. Prolonged taking of temazepam eventually results in it being less effective. Some studies have observed this to occur after as little as one week's use, while other studies have said that it is not really possible to become tolerant of the drug.

Temazepam withdrawal symptoms are like those experienced by alcoholics and barbiturate users; the higher the dose and the longer the drug is taken, the greater the risk of these being unpleasant. Among the so-called benzodiazepine drugs, temazepam had the highest rate of overdose problems, according to one analysis. In 1993, a UK study found temazepam to have the highest number of deaths per million prescriptions among medications commonly prescribed. It is a drug with a high potential for misuse. Some of those who have been prescribed temazepam by their doctor have used it for other purposes, and one such case was that of Linda Lees.

5.2 ATTEMPTED MURDER, HELSTON, CORNWALL, 2008

When Linda Lees discovered that her husband Paul, who was in the Royal Navy, had left her for another woman, she planned to murder him and at the same time she would commit suicide. This she would do in their garage and by means of carbon monoxide from her car exhaust. With the help of temazepam, which she was taking for her depression, she would render him unconscious while in her car, and then drive into the garage and arrange for the car exhaust to be vented into the car interior. That was the plan.

This was a crime of passion on her part, said the detective who investigated what had happened at their home in Cornwall in 2007. However, the situation which transpired was described later by the judge as "a very bad detective novel." It was more of a farce.

In 2007, Linda Lees was 45 and lived in Bosnoweth road in Helston, Cornwall, in the house that had previously been the family home. There she lived with her husband Paul, who was

43 and was employed as an avionic instructor at HMS Sultan.[‡] Linda, meanwhile, ran a car-hire company in Exeter. They had first started going out together when they were both teenagers, and they had married in March 1985. They met when Paul was stationed at the Royal Naval Air base at Culdrose, very near Helston.

Paul had confessed to having had a previous relationship with a woman in the Women's Royal Navy Service, popularly known as WRENs. Nevertheless, he and Linda remained married for almost 23 years and they had a 21-year-old daughter, Hannah, who was born in 1987.

It was in the nature of Paul's job that he spent quite long periods away from home, and it was during one of these that he had met a woman in Baltimore. They began an affair, which included a visit together to New York. When he returned to the UK, she came back with him and they bought a new semi-detached house in the Southsea area of Portsmouth. Linda, meanwhile, had also found consolation in the arms of a neighbour, Warren Sibbald, who lived only four doors away and whom she had known as a friend of Paul's.

In October 2007, Paul told his wife that he wanted a divorce, and he arranged to return to Helston to discuss the matter and collect his belongings. Linda agreed, but she had a different plan; she would drug Paul with her antidepressant pills and then make it appear that they had committed suicide together. Linda had become depressed by her situation and twice she had to go into hospital in March 2008 because she had taken a combination of paracetamol and temazepam. In future, she would give wayward Paul this drug instead. However, it was not going to be easy to do, as she was to discover.

Between Easter and July, Paul met up with Linda three times, and on each occasion, she tried to carry out her plan when they dined together. However, it would appear that in none of her attempts did she give Paul a fatal dose of temazepam, although each time she increased the amount of the drug she put in his food and drink.

[‡]HMS Sultan is a Royal Navy training establishment which specialises in marine engineering, defence, and survival techniques.

The first time she attempted to put her plan into action was at their Helston home on Tuesday 25 March 2008. It involved a takeaway, which Linda had ordered, and to which she added temazepam to the part that she served to Paul, and as they ate it, they watched TV.

Later, he said that the wine she served with the meal tasted salty, but he didn't remember much of that evening except that he awoke to find himself naked with his wife wearing surgical gloves and shouting at him. He fled to the bathroom and locked himself in, but when he emerged Linda was waiting for him and she pushed him down the stairs. Paul was not badly injured but rang 999 and spoke to the police. They duly arrived only to be told that it was an accident. Linda said she had been trying to commit suicide and had also given temazepam to Paul to make him pass out so he would not stop her. The police arranged for Paul to give a sample of blood and this indeed indicated he had the drug in his system. Linda later said that she had given Paul temazepam on this occasion because she wanted to gain access to his mobile phone, hoping to learn more about other women in Paul's life.

The second time they ate together was the following month at a Chinese restaurant in Exeter. Ostensibly, she had planned an intimate meal when she hoped they would become reconciled. It soon became obvious that was not going to happen so she intercepted the waiter before he could bring the meal Paul had ordered to their table. She told the waiter that they were having a romantic evening to celebrate their reconciliation and she had brought her wedding ring, and to give Paul a surprise, she wanted to hide it in his curry. The waiter agreed, and it gave Linda the opportunity to add temazepam instead. After they had eaten the meal, Paul said it made him feel somewhat groggy but not enough to prevent his driving to their house in Helston while Linda offered him some bottled water to which she had added more temazepam. By now he was very groggy and fell deeply asleep. There was no way she could carry him into her car, so all Linda could then do was just let him sleep it off, which he did.

The third time they met, on Sunday July 20 2008, Linda again ordered a takeaway although Paul's visit was not a social occasion, as the previous get-togethers had been. This time he had just come to collect more of his belongings but she later claimed that they had sex together. She added a much larger dose of the drug to his meal and, as a result, he became very disorientated. He allowed Linda to take him to her car, thinking she was going to drive him to get medical help. Instead, she drove him around Helston until he became unconscious.

Then she drove the car into the garage at the house where she attached a hosepipe to the car exhaust and, with both of them sitting in the car, she turned the engine on, thinking that they would soon die due to carbon monoxide poisoning. It would appear as if they had had a suicide pact. What Linda did not realise was that, while this method of committing suicide was common in earlier times, the introduction of catalytic convertors in the mid-1990s now removed this toxic gas from engine emissions. After a while Linda realised that nothing was happening, so she got out of the car and went indoors. The plan had failed. She told her neighbour Warren what had happened and he rang the police.

Forensic examination of both of them revealed no carbon monoxide in their blood.

Linda pleaded guilty at Truro crown court in October 2008 and was given a 49-week prison sentence, suspended for 18 months, and ordered to pay £300 court costs. It was Paul who pleaded for her not to be sent to prison because he said their daughter needed her mother. The judge agreed but he instructed Linda not to visit her husband's home in Southsea or his place of work. She was never to make contact with him again, except through her solicitor.

5.3 MURDER, ABERDEENSHIRE, 1994

Few murderers these days justify a whole book devoted to them and their crimes, but such a one is Malcolm Webster.[§] He killed his first wife in Aberdeenshire by dosing her with temazepam to make her unconscious, then crashing their car and setting it alight, so burning her to death. He tried to murder his second wife in New Zealand the same way but thankfully she survived. His motives? Each time he planned to cash in on their life

[§] *The Black Widower: the life and crimes of a sociopathic killer*, by Charles Lavery, published in 2012.

insurance to pay for his reckless spending sprees. This had been a successful way of solving his financial problems when he murdered his first wife and collected £200 000. With his second wife, he planned for a massive NZ$1.9 million insurance pay-out (equivalent to £700 000), but thankfully it failed and she lived to testify against him in court in Glasgow.

Malcolm Webster was born in Guildford, Surrey, on 18 April 1959; one of twins, the other being a girl, Caroline. They also had an older brother, Ian. Webster's father was Detective Chief Superintendent Alexander "Sandy" Webster, who was head of the Fraud Squad of the Metropolitan Police in London. Webster's mother had been trained as a nurse, but she was a shy individual who had few friends.

Webster junior was a somewhat wayward child. At school he was a loner and was clearly fascinated by fire, to the extent that his playground nickname was "pyro." He had discovered that nail varnish and nail varnish remover were flammable – the former contains **nitrocellulose** and the organic solvent **ethyl acetate**, the latter is a mixture of ethyl acetate and **isopropyl alcohol**. He was to use this knowledge in various arson attacks later in life. Webster left school at 15, and then had a succession of low paid jobs working as a driver and an office clerk before deciding that a career as a nurse was the best way for him to earn a living. Nor did he find it difficult to attract women; he was good-looking and charming.

In 1978, when he was 19, he got a job in a nursing home where he began to steal cash and valuables from the residents. He seduced the 15-year-old daughter of the owner and made her pregnant, causing her to have an abortion. He was sacked, but the police were not informed. Nevertheless he found other nursing jobs and continued to live beyond his means.

In 1993, Webster met Claire Morris at a party in London. She was from Scotland and apparently it was love at first sight. She was also a trained nurse and when Webster got a job at Aberdeen Royal Infirmary, Claire agreed to move to Scotland with him, and there she registered for a BSc in health care at the University. They got married on 3 September 1993 in the King's College Chapel in Aberdeen in an elaborate affair which many of their friends and family attended. They moved into their new home, Easter Cattie Cottage, near Oldmeldrum, a small village about

17 miles from Aberdeen and where Claire Morris was originally from. Soon after their marriage, Claire began to feel very tired on some days and that was due to Webster secretly adding diazepam and temazepam to her food and drink. At weekends, she would sometimes sleep for more than a day. Webster had a history of going to his doctor claiming to be suffering from conditions for which diazepam was prescribed, and he also had unrestricted access to temazepam at Aberdeen Royal Infirmary.

Webster was always in financial difficulties because he spent lavishly in up-market shops like Harrods when he was in London. He was heavily in debt but he had a plan to solve his problems. He would insure his 32-year-old wife and kill her. He persuaded her to take out life insurance amounting to £200 000.

His plan was to drug her with temazepam, involve her in a car crash which would cause the car to burst into flames, and this would not only ensure she died but that there would be nothing left of her that would show that she had been drugged. The plan worked perfectly. He collected the insurance and he also applied for a widower's pension, so securing extra payments of £10 439 per year when in fact he was not eligible to receive it.

The supposed accident occurred on the road between Auchenhuive and Tarves Road in Aberdeenshire on the night of 27/28 May 1994, when Webster claimed he had suddenly had to swerve off the road to avoid colliding with a speeding motorcycle. In the front seat of the car was a drugged Claire. It was not until his trial many years later that it was proved that she had been given temazepam, which was detected by pathologist James Grieve in a tiny portion of her liver which had survived the fire.

The car was a Daihatsu Sportrak four-wheel drive and it appeared to have left the road and hit a tree. Webster got out of the car and took newspapers out of the boot, soaked them in petrol, and placed them in the car and in the engine compartment. Just then a coach driver stopped his vehicle and asked if Webster needed help. He said he was OK and denied there was anyone else in the car, so the coach drove off. Other people stopped their cars and offered help but Webster told them he was alone and that things were under control. One of them went to look for a local farmhouse to ring for assistance.

Then another car pulled up and a woman got out. She could actually see Claire slumped over in the front seat, but as she approached the car, its engine started to give off smoke and then the whole car burst into flames. She was horrified to see them envelop Claire whom they eventually reduced to a charred mass. (At the time she was identified from her dental records.) It is to be hoped that she was already dead when the fire started, murdered by an overdose of temazepam.

By now an ambulance had arrived, to be followed eventually by the fire brigade. Webster was taken to the Royal Infirmary where, at 4 am, he was examined and found to be totally unharmed although he complained mainly of neck pain due to whiplash. He was admitted to the hospital where he stayed for a week, during which time he organised Claire's funeral, and this he attended wearing a neck brace and with his arm in a sling.

A police officer who attended the crash scene was puzzled by what he found and continued to investigate the supposed accident in his spare time even after it was ruled to be genuine and the case closed. Why had Claire not escaped from the crashed car? And why were there no skid marks at the site of the crash as would normally be expected of a car trying to avoid a collision with another vehicle. Nor did the car seem to have been damaged in any way. This seemed to indicate that the car was not speeding and yet the crash was supposed to have been so bad that Claire had been rendered unconscious.

One of the firemen who attended the scene also had doubts about what had happened. Why didn't Webster pull his wife out of the car? And why should the car catch fire; that just never happened in a case like this. Rather strangely, among the wreckage was a petrol can behind the driver's seat. It all seemed very odd but no further action was taken, partly because Webster's father used his influence to have the case closed. Webster collected the £200 000 insurance due on Claire's life.

Meanwhile, he had found consolation with other women in the weeks after Claire's death. One was Caroline McIntosh, whom he had met at Aberdeen Royal Infirmary. He eventually took her on a trip on his new yacht, and the couple stayed at a hotel in Inverness. She was impressed by his brand new Range Rover car. He invited her to stay the night at his home, East Cattie Cottage in Oldmeldrum, which she did, and so began a relationship that was to be renewed whenever he was in Aberdeen.

The insurance money following Claire's death was all gone within six months, none of which appeared to have been used to clear the debt on his credit cards, of which he had about a dozen.

5.4 ATTEMPTED MURDER, NEW ZEALAND, 1998

In December 1994, Webster got a job in Saudi Arabia selling specialist medical computer software. One night in Riyadh, in May 1996, he met his next intended victim at a dinner party. She was Felicity Drumm, and for her it was love at first sight. She worked at the King Fahd Hospital and had managed to earn enough to not only pay off the mortgage on her home back in Auckland, New Zealand, but to have a healthy bank balance of NZ$98 000 (around £50 000). Webster proposed to her and she gladly agreed to marry him.

Webster worked at the Tawam Children's Hospital in Abu Dhabi, but now it was essential for him to leave because of the rather strange deaths of three very young children who had apparently succumbed to heart failure, which is almost unheard of in children so young. When each one died, it was always while Webster was on duty and it is now believed that he had injected them with excess insulin, maybe seeing how this might work as a way to kill someone.[¶] The children did not undergo autopsies and their real cause of death was never discovered. In any event, Webster didn't face charges, again possibly thanks to his father's intervention with the Foreign Office, but he had to leave the country immediately and did so.

Webster found a new job at his old hospital in Aberdeen, and in May 1997, he and his new wife flew to Scotland and moved in to Easter Letter Cottage, at Lyne of Skene, Aberdeenshire. Felicity was now pregnant, and baby Edward was born in May 1998. However, Webster still planned to murder her in the same way that he had killed his first wife, but that "accident" would take place in New Zealand. The family planned to move to Auckland, but not before some strange happenings in Scotland.

[¶]The first person to murder someone by insulin was Bradford nurse Kenneth Barlow, who killed his wife this way in 1957. He was found guilty and imprisoned. He was released in 1984 when he was 65 years old.

In September 1997, there was a fire at their cottage and they lost many of their belongings, although Webster was able to claim on insurance. Just before they moved back to New Zealand, they put the rest of their things into storage at a facility in Aberdeen which also burned down in a fire that did £5 million of damage. Webster had visited the storage depot just before the fire started, which was eventually blamed on a workman's blow torch being left on in the roof area. However, it now looks likely that it started in Webster's storage space, which was directly below that part of the roof area. Webster was able to claim £68 000 from the insurance, although he first said his loss was £87 000, which the insurance firm disputed.

In the autumn of 1998, they finally moved back to New Zealand where they stayed with Felicity's parents while buying a new house, which was also mysteriously damaged by fire one evening; as was her parent's home when an armchair caught fire while they were all in bed. It was easily extinguished. In December 1998, Felicity had transferred her savings of NZ$140 000 to what she assumed was a joint account with Webster.

All was now set for a repeat of the murder he had successfully got away with in Scotland in 1994. One weekend, the family set off for what Felicity thought was to be a picnic on the coast but the car ended up in a forest area. By then, she was comatose in the car due to her having been given excess temazepam that morning, but before Webster could set it alight, he had to take his son well away from the vehicle, and while he was doing this, Felicity's mobile phone rang persistently and woke her up. It was her father who had just opened a letter addressed to his daughter marked "urgent" and discovered that her bank account had been cleaned out. He told her to get home as quickly as possible.

Now things began to unravel for Webster. Felicity's father later discovered that the car boot was full of crumpled newspapers. When doused with petrol, they would have burned quickly and fiercely. Webster realised the game was up, packed his bags and fled the country.

5.5 PLANNED MURDER, OBAN, SCOTLAND, 2006

Webster returned to Scotland and now had a new partner, 41-year-old Simone Banarjee. They lived in Oban on the west coast, and she was employed as an operating theatre manager at the local hospital where Webster also worked as nurse, specialising in the careful lifting and moving of patients. He proposed marriage to her even though he was still married to Felicity.

Simone was earning a good salary but more important from Webster's point of view was that her family had set up a trust fund for her. He persuaded her to sign over her life insurance policies to him, and he did the same for her. They had bought a yacht together called the *Nina* and were planning to take part in a transatlantic race from the Canaries to the Caribbean.

Then, in December 2005, Webster rang Simone to say he was terminally ill with cancer. He had gone to Kent to attend his father's funeral and said he had been to the Royal Marsden Hospital where his condition was diagnosed as chronic lymphocytic leukaemia. Simone travelled to London and discovered he was now bald, as might be expected due to chemotherapy. She was now desperate to be married to him. In February 2006, Simone changed her will leaving all to Webster (more than £300 000) and was rewarded in September with an engagement ring worth £6000. Luckily for her, the police were now on his trail.

Simone Banarjee was eventually persuaded by the police that Webster was going to kill her once he'd robbed her of all her money and that this would happen when they went sailing together on his yacht. It appears he planned to do this by drowning her and he had punctured her life jacket, something she discovered later when she compared her jacket with others. That was the end of her affair with Webster, who left Oban immediately. Simone Banarjee's house was eventually searched by police after they received information that Webster had embezzled funds from a local angling club. During the search, the police seized a stolen laptop and an unlicensed gun, which Webster later claimed was an antique.

Webster was unaware the police now had a file on him and had launched an investigation called "Operation Field," which included a re-examination of his first wife's death. The investigation into Webster took five years and involved 1000 people being interviewed. He was consequently charged in 2009 with the murder of Claire Morris, the attempted murder of Felicity Drumm, and attempting to bigamously marry Simone Banarjee in order to gain access to her estate. He was also charged with various offences: arson, selling drugs, and illegally poisoning his wife.

The prosecution accused Webster of putting a container of petrol, newspapers, and a lighter into a vehicle and appearing to crash into a tree in Auckland, on 12 February 1999, in an attempt to murder his passenger, Felicity. What might have happened to her was graphically illustrated with pictures of what had happened to his first wife in Scotland. Only the jury were shown pictures of Claire's charred remains, since they were regarded as too disturbing to show to others in the court that day.

Webster was convicted at the High Court of Judiciary in Glasgow on 19 May 2011 after the longest ever criminal trial in Scotland that involved only one accused person. It had begun on 1 February 2011. He was found guilty by a jury of nine women and six men and sentenced to life imprisonment on 5 July 2011, with a minimum term of 30 years.

5.6 MURDER, GENEVA, ILLINOIS, 2016

It seems that Julia Gutierrez successfully disposed of her husband Eduardo with temazepam in January 2016. She had made a fruit smoothie for him and had added the contents of several temazepam capsules to it. Soon after drinking it, he passed out. She then took an overdose of the same drug, seemingly in order to kill herself, but she survived and the police were called. She was subsequently arrested.

Julia and Eduardo lived in a detached house on the corner of Crissey Avenue and Oak Street in a leafy suburb of Geneva, Kane County, Illinois. The house is estimated to be worth around $400 000. The couple had been married for 31 years but all was not well.

Julia and her husband were both 53 and were regarded by their neighbours as nice people, although they had not many friends; he was a bus driver and she was self-employed. She did a variety of jobs, such as personal assistant, and she had been working for 15 years, earning around $8 000 a year. When her finances were examined, it appeared she was in debt to the local hospital and an ambulance company, and there were bills for jobs that had been done on the house and which had not yet been paid.

However, it appears she had two bank accounts with balances of around $110 000 as well as $50 000 in an investment account. At the time of her husband's death, she had sent a package to a friend with a note saying she was going to take her own life.

This was not the first time that the police had been called to their home. In 2002, they were called when Eduardo was taken ill as a result of drinking a milkshake to which temazepam had been added, but no action was taken because Eduardo said he would not press charges against his wife. However, Julia's sister Rachel Mooney claimed later that it was an attempted murder. The case is ongoing.

CHAPTER 6

Potassium Chloride: Essential to Life Yet Deadly

A word in **bold** *indicates that further information can be found in the Glossary. Only the first time the word appears in a chapter will it be so indicated.*

We all know that salt is sodium chloride and that too much salt in our diet is dangerous because it can be a contributing factor to high blood pressure. We take a different attitude when it comes to potassium chloride, which is chemically very similar, and some people are advised to use a product called LoSalt, which is two thirds potassium chloride, when cooking and seasoning food. However, when it comes to injecting potassium chloride, we have a very dangerous substance which can kill within minutes. So, what is so special about potassium that makes it both essential to life yet deadly?

Potassium exists in Nature as the positive ion K^+ and consequently needs a counterbalancing negative ion such as chloride Cl^-, as in potassium chloride, KCl. The same need for a negative counterpart also applies to sodium, and common salt is sodium chloride, NaCl. The chemical symbols for these key elements come from the Latin names for potash,[†] which was

[†]Potassium carbonate K_2CO_3.

More Molecules of Murder
By John Emsley

Published by the Royal Society of Chemistry, www.rsc.org

kalim, and soda,[‡] which was *natrium*. Both these elements are abundant on this planet and they are present in seawater; sodium to the extent of 10.8 grams per litre and potassium 0.4 grams per litre. Not surprisingly, they play a vital role in all living things.

What is perhaps less appreciated is that an isotope of potassium is radioactive, and 4000 potassium atoms undergo radioactive decay every *second* in the average human. Potassium decays by emitting a **beta particle** (a negative electron) from the nucleus and thereby goes from element atomic number 19 to element atomic number 20 and that is the gas argon. This radioactivity is not the reason why potassium can pose a threat to health because this kind of radiation is relatively harmless.

Potassium chloride is widely used to treat hypokalaemia, the condition when potassium levels in the body are too low, as can happen with seriously ill patients, who then face the possibility of a heart attack. It is for this reason that hospitals stock 10 mL vials of concentrated potassium chloride solution, and care must be taken when these are used to ensure that the solution is diluted to the right strength. Others have misused these vital medical supplies for their own depraved motives.

6.1 POTASSIUM, THE VITAL ELEMENT

There are three aspects of potassium which are important: fertilizer, food, and function.

All plants absorb potassium from the soil, and whatever crop a farmer grows, it will remove potassium from the land, and this must be replaced if future yields are not to suffer, which is why this element is one of the basic ingredients of fertilizers. All fertilizers should contain three essential elements which crops need, and they are nitrogen, phosphorus, and potassium, shown as the letters 'NPK' on fertilizer sacks.

The potassium for fertilizers comes from mineral deposits, such as sylvite, which is mainly potassium chloride, and carnallite, which is potassium magnesium chloride. The sylvite deposit underneath the North York Moors National Park is vast and capable of supplying much of the UK's needs for many years

[‡]Sodium carbonate Na_2CO_3.

Figure 6.1 Potassium chloride crystals. © Shutterstock.

to come. The mined mineral will be transported along a 23-mile underground tunnel to its processing plant outside the park in order to avoid any impact on the park itself.

Potassium chloride (Figure 6.1) itself is not a particularly soluble salt, but enough does dissolve in the soil water to supply the needs of plants. The potassium ions are used by a plant in lots of ways and they activate at least 60 different enzymes and, in particular, the ones which govern photosynthesis. Another important role is their control of the stomata in leaves. These are the pores through which a plant "breathes," taking in carbon dioxide from the atmosphere and releasing oxygen gas.[§] The stomata swell up with water and this allows them to open and function as required, but if there is not enough K^+ then they close, and this is one way that plants can survive a drought when its roots cannot absorb K^+ through lack of water in the soil. Potassium is also needed for making and transporting sugars around the plant, and in forming proteins. If there is not enough potassium in the soil, and even if other nutrients and water are abundant, then a plant will struggle to grow and may be unable to produce all the components that are required if we are to consume it as food.

[§]This is the reverse process to the one that occurs in our lungs; we breathe in O_2 and breathe out CO_2.

Table 6.1 Potassium in common fruits and vegetables.

Fruit	mg per 100 g[a]	Vegetables	mg per 100 g
Bananas	358	Spinach	558
Kiwi	316	Potatoes	535
Melon (cantaloupe)	267	Mushrooms	448
Cherries	222	Brussel sprouts	398
Grapes	191	Parsnips	375
Peaches	190	Beetroot	325
Oranges	181	Carrots	320
Strawberries	153	Broccoli	316
Apples	120	Cauliflower	299
Pears	116	Celery	284
Pineapples	110	Sweetcorn	294

[a]100 mg per 100 g is 0.100%.

Virtually all foods contain potassium, the exceptions being things like vegetable oils, butter, and sugar. Some foods are particularly rich in potassium: seeds and nuts may have up to 1% by weight, compared with a more normal range of 0.1–0.4%. Some foods have more than 1%, such as the cereal All-Bran with 1.1%, butter beans with 1.7%, Marmite with 2.7%, and instant coffee granules, which have the most at 4.0%. Among fruits, bananas are the top end of the range with almost 0.4%, whereas pears and pineapple are the lower end and have only 0.1%. Vegetables tend to have more potassium that fruit, with potatoes having more than most with 0.5% while sweetcorn comes at the lower end, but even that has 0.3%. Other examples are shown in Table 6.1.

Dried fruits have much higher levels of potassium, with dried apricots being the highest at 1160 mg per 100 g. Other dried fruits are raisins with 750, prunes 730, dates 700, and figs 680 mg per 100 g.

6.2 POTASSIUM IN THE BODY IN SICKNESS AND IN HEALTH

In an ideal world, the average adult woman would weigh 70 kg (11 stone) and the average man would weigh 80 kg (12.5 stone), and they will contain about 120 g of potassium, although this can vary between 110 and 140 grams depending on the amount of muscle in their body; the more muscle there is, the more potassium the body needs.

A normal person needs an intake of 3.5 grams of potassium a day, and most of this comes from the cereals, fruit, and vegetables we eat. This is much more than the basic 1.5 grams a day intake of sodium which we need.¶ There is nowhere for the body to store potassium and we lose about the same amount of potassium a day, but a normal diet will provide all we require. In reality, we need a constant throughput of this essential element to make lean tissue and keep the kidneys working.

Potassium is found in all parts of the body. Red blood cells have most, followed by muscles and brain tissue. Its highest concentrations are inside cells. Potassium has other functions in the human body, such as regulating the fluid levels within cells and enabling proteins to become soluble and mobile. Muscles need potassium if they are to function, which is why this element is vital to the beating of the heart. Too little potassium in the body and we feel weak, too much and we are in danger of stopping the heart.

So what metabolic role do potassium ions perform in our bodies? The chief answer is that they are part of the mechanism by which messages are transmitted along our nerves and in and out of our muscles. Cells contain potassium and this can move in and out of the cell through millions of tiny channels in the cell membrane. As many as 200 potassium ions per second can move through each channel and, as they do, they create a wave effect which is like an electric current passing down the nerve. The bite of the black mamba snake kills its prey by injecting a chemical which blocks the potassium channels.

Certain kinds of diuretics can result in a loss of potassium and so deplete the body of this element. When this treatment is prescribed then levels of potassium need to be monitored and if necessary supplemented by prescribing tablets to replenish it. However, ill health due to potassium deficiency is rare, because it is almost impossible to avoid potassium if we eat a normal diet and aim to consume the five portions a day of fruit and vegetables recommended for healthy living.

¶Daily dietary intake of sodium is generally double this amount, and even more in some individuals. The UK Government's Scientific Advisory Committee on Nutrition (SACN) published a comprehensive report on this, see *SACN Salt and Health Report*, in August 2003, which can be downloaded from the website www.gov.uk/government/publications/sacn-salt-and-health-report.

Potassium salts are an essential part of post-operative treatments and are given in tablet form rather than injections. However, some treatment requires better control of this vital element, which is why hospitals stock concentrated solutions of potassium chloride in the form of 10 mL ampoules which contain 1.5 g of KCl in 10 mL of water, and this must be diluted to 500 mL with water or sodium chloride solution before use.[‖] The ampoules are very similar to those which contain sodium chloride, and this has caused nurses in the past to mistake them, with fatal consequences for the patient. Hospitals have introduced strict clinical governance measures to ensure safe storage, prescription, and delivery of KCl solution so that this cannot happen.

6.3 POISONOUS POTASSIUM

As some of you may know, potassium chloride is one of the chemicals used to execute prisoners in certain states of the USA. An intravenous injection of concentrated potassium chloride solution will kill someone within minutes. This is done legally in some states, and in the past, if a condemned person agreed to donate their organs for transplant surgery then they could be executed by what is described rather paradoxically as a "nontoxic lethal injection" of potassium chloride.

Potassium offers a means of murdering someone provided you can inject them with it, and that may be possible if they are in a situation where they might expect to be given injections. What also aids the would-be murderer is that potassium is not something that an autopsy would reveal unless the pathologist was specially requested to look for it. As we shall see, this combination of conditions has allowed some serial killers to remain undiscovered for years.

Excess potassium in the blood can be fatal because it interferes with potassium ions moving in and out of cells and in relatively small doses it can mess up the central nervous system, causing convulsions, diarrhoea, and kidney failure. A very large dose, such as a solution containing around three grams of KCl,

[‖] Another form of potassium is Addiphos, which is a mixture of potassium dihydrogen phosphate (KH_2PO_4) and disodium hydrogen phosphate (Na_2HPO_4) with a little potassium hydroxide to ensure its pH is not acidic.

will interfere with the action of the heart muscle and cause it to stop beating within a few minutes, resulting in a rapid death.

Injecting a solution of potassium chloride into the bloodstream means there is too much potassium on the outside of cells, and this blocks the transfer of potassium across the cell membrane. This in turn stops the local activity in the cell wall and so rapidly stops the working of the heart muscle. This method of ending someone's life is not going to be easily detected, which is something that a few murderers have relied upon as a way to escape detection. Simply consuming lot of potassium chloride by mouth, *i.e.* enterally, will not have the same result, which is why we can safely buy products such as Lo-Salt. Of course, if a person were to consume a lot of this product at one time, say 20 g or more, then death might result and there have been cases of this happening.

The eminent toxicologist Charles Wetli, of Alpine, New Jersey, reported in 1978 the case of a medical secretary who poisoned herself accidentally with potassium chloride. She went on a liquid protein diet and had been prescribed slow-release potassium chloride pills because such a diet can lead to a condition known as hypokalaemia.** You can live with this and not notice its effect unless you engage in strenuous exercise. However, when living on a diet which excludes foods that are rich in potassium, then the body's supply of this vital element can be depleted, and this is what happened to the medical secretary. She took potassium pills whenever she felt tired, and on the day she died, she took more than 40 of them. She complained of severe diarrhoea but was told that this would clear the excess potassium from her body, but that night she died. Her blood showed a gross excess of potassium and this was the cause of her death.

Potassium chloride has been used to commit suicide but not always successfully, and an example of this involved a 20-year-old nurse in Toulouse, France. She had injected herself with four hypodermic doses, each of 20 mL of a 10% solution of KCl. This caused a heart attack and she was then treated for this condition

---

**This is defined in terms of the level of potassium in blood being below 3.5 millimoles per litre, which is 136 mg per litre. The normal level for potassium in the blood is between 4.0 and 5.0 millimoles per litre (156–195 mg per litre).

and survived. A tube was inserted into her windpipe and oxygen-enriched air pumped into her lungs while the medical team dealt with the poisoning, which they did with an intravenous solution of 1% sodium chloride and an injection of adrenaline, along with insulin and glucose, followed by a solution of sodium bicarbonate. After 20 minutes, the nurse showed clear signs of recovery although she became delirious and was eventually hospitalised in a psychiatric ward.

When it comes to using potassium as a murder weapon, it seems to have been mainly the choice of women and those who work as nurses; no doubt because only they have easy access to it. Here, we look at one nurse who used it to murder her child, one nurse who used it to murder lots of other people's children, and one who used it to murder elderly people. Then, we will examine a case of the family that was eliminated with it because a religious cult thought the father was a threat.

6.4 INFANTICIDE AND EXECUTION, SHERWOOD, ARKANSAS, 1997

Christina Riggs was a 26-year-old nurse living in Sherwood, Arkansas, with her two children, 5-year-old Justin Thomas and 2-year-old Shelby Alexis. By reason of her occupation, she had access to potassium chloride and hypodermic syringes, so when she decided to murder her children and commit suicide, she chose this chemical as the preferred method. However, when she attempted to carry out her plan, she bungled it when it came to killing herself with an injection of potassium chloride – although when it was her turn to be executed a year later, the same chemical caused her death within minutes.

Ms Riggs was born in Lawton, Oklahoma, in 1971, and she lived most of her life in Oklahoma City, where she liked a good time, smoked cigarettes and marijuana, and drank heavily. She was free with her favours and when she was 16, and still at school, she had a baby boy whom she handed over for adoption. When she left school, she trained as a nurse and got a job at the local Veterans Hospital. She also worked part-time in a care home. She had a steady boyfriend, Timothy Thompson, but when she became pregnant, he realised it was not his and he left her. Baby Justin was born in June 1992.

Soon, she found another partner, John Riggs. They married in July 1993 and their daughter Shelby was born in December 1994. After her birth, Christina returned to working as a nurse, now at the Baptist Hospital. Things were far from perfect at home and their marriage was under stress, due in part to her boy Justin who suffered from ADHD (attention deficit hyperactivity disorder). John Riggs found his step-son difficult to cope with, and one day he hit the child so violently that the boy required medical attention. Eventually, John left Christina and the couple were soon divorced. She and the children moved to Sherwood, a town of 30 000 in Pulaski County, Arkansas, and there she got a job at the Arkansas Heart Hospital. Her mother also moved so as to be close to her daughter.

By now Riggs was obese and weighed 20 stone (280 pounds) but was earning a respectable $17 000 a year as a nurse and had maintenance payments from her divorced husband. Her mother helped her by often looking after the children. Clearly it was not financial pressures which drove Riggs to doing what she did, although she later maintained that this was so. Part of the reason for claiming to be short of money was that she continued to live the lifestyle she had enjoyed when she was younger. She would sometimes put the children to bed and, when they were asleep, she would leave the house and go to a karaoke bar for the evening. However, clearly this was not the life she had dreamed of when she was a young girl and it appears she decided to end it, not only for herself, but for the children as well.

On the evening of 4 November 1997, she gave the children the drug Elavil, which is **amitriptyline**, and this is used to treat various mental conditions. This sent them into a deep sleep. At 10 pm, she took a hypodermic syringe and charged it with the potassium chloride solution she had stolen from the hospital, and injected Justin into his neck. This caused him so much pain that he woke up and started screaming. Christina rocked him back to sleep while he died and then, just to make sure he was dead, she pressed a pillow over his face. She smothered Shelby without injecting her, then placed the two children side-by-side in bed. In order to kill herself painlessly, she then took 28 of the Elavil pills before injecting potassium chloride solution. This was to no avail as she was so overweight that she was unable to find a vein in her arm for the injection and the fluid she did inject did not circulate as she thought it would. Nevertheless, she passed out.

The following day, her mother went round to the house but found the door locked and there was no response when she tried to contact her daughter. So, she rang the police and when they broke in, they found the two dead infants with Christina unconscious on the floor beside the bed. She was rushed to hospital where doctors saved her life. As soon as she was able to be discharged from their care she was arrested.

In 1998, she was put on trial for murder and entered a plea of not guilty because of mental illness, but she refused to put up a defence, although a psychiatrist testified to the court that she was mentally ill. That testimony had no effect and Christina admitted to the court that she had killed her children and said she wanted to die as well. The jury respected her wishes and duly found her guilty. A sentence of death was passed by the judge.

For a year, Christina languished in jail and managed to increase her weight to 22 stone (310 pounds). Then, on Sunday 30 April 2000, she was flown from Newport prison to the Cummins Unit in Arkansas for her execution on Tuesday 2 April. She would be the first woman to be executed in Arkansas since 1845. It took place at 9:30 in the evening and she was given an injection of potassium chloride solution. By 9:40 she was dead. Her last words were "I love you, my babies." Perhaps, in her own misguided manner, she really did.

6.5 MULTIPLE CHILD MURDERS, GRANTHAM, ENGLAND, 1991

Another murderer of young children was also a nurse, albeit only a trainee one, and she was 23-year-old Beverley Allitt. However, it was not her own off-spring that she targeted but those who had been admitted to the hospital where she worked, which was the Grantham and Kesteven General Hospital in the town of Grantham, Lincolnshire, 24 miles east of Nottingham. In 1991, she murdered four of her charges by injecting them with potassium chloride solution; the youngest was a seven-week-old baby boy, Liam Taylor, and the oldest was an 11-year-old boy, Timothy Hardwick. She also attacked six others during her brief time at the hospital. It appears that Allitt enjoyed the excitement that ensued as doctors and nurses fought to save a life, and in which she played an active role.

Her first victim was baby Liam, who had been admitted to hospital on 21 February, only a couple of weeks after Allitt joined the staff on a temporary basis – she had yet to qualify as a nurse but the hospital was short-staffed. Liam was suffering from congestion of the lungs, but suddenly took a turn for the worse and was then rushed to emergency care where he started to recover. His parents stayed at the hospital the following night and slept in a special bedroom reserved for parents whose children were dangerously ill. Allitt was very sympathetic to their plight and even volunteered to work a night shift in case baby Liam underwent another emergency, which is just what happened and his heart stopped beating. Although doctors struggled to keep him alive, he died in the early hours of the morning.

Allitt's next victim died only two weeks later on 5 March and he was Timothy Hardwick, who suffered from cerebral palsy and was admitted to the hospital following a critical epileptic fit. Suddenly, his heart stopped beating and Allitt, who was the sole nurse attending to him, called for help but it was already too late to save him despite the efforts of a doctor. No cause for his sudden death was discovered, despite a post-mortem being carried out, and it was attributed simply to epilepsy.

Her next attack with potassium chloride failed to kill the one-year-old girl she injected, whose life was saved by doctors and who then transferred her to Nottingham University Hospital for more specialist treatment. While she was there, a doctor noticed a suspicious-looking injection site under the child's armpit, but this was not followed up because the baby recovered. She next injected another baby, this time with insulin, and he too survived after being rushed to Nottingham and there the doctors did note that his blood sugar level was dangerously low.

Clearly, insulin was not as effective as potassium chloride in creating the crisis situation she intended, so it was back to using potassium chloride again. Her next victim, a five-year-old boy, survived two of Allitt's injections before he too was moved to Nottingham – and survived – and the same thing happened with two-year-old Yik Hung Cha who had been admitted to hospital after falling and fracturing his skull. He too survived.

Next, it was the turn of prematurely-born twins, one of whom died while under Allitt's care; the other was rushed to Nottingham and survived. The parents were so impressed with the strenuous efforts which Allitt made to save their baby's life that they asked her to be the child's godmother when it was christened the Sunday following its return home. Allitt was suitably contrite over the couple's recent tragedy and cynically accepted the role.

The nursing staff at Grantham and Kesteven Hospital made strenuous efforts at hygiene in the belief that the children's ward was infected by a deadly virus. The ward was cleared and decontaminated. All to no effect. Suspicion then began to fall on the nurse who was most often involved when these various calamities had occurred and a special watch was made of Allitt's movements.

She then attacked 15-month-old Claire Peck, who had been admitted due to a severe asthma attack and who was breathing with the help of a tube. Allitt was the nurse assigned to the ward and she was the only one present when the girl suddenly suffered cardiac arrest. Thankfully, an emergency team was able to save her. No sooner had they left the ward than the child had another attack and this time they could not save her. An autopsy revealed that death was due to potassium chloride poisoning. The police were called and Allitt was arrested. She was brought to trial in March 1993 and found guilty of multiple murders, and given 13 life sentences and committed to Rampton Secure Hospital, from which she is unlikely ever to be released.

6.6 SERIAL MURDER, LUGO, ITALY, 2014

Another nurse who resorted to potassium chloride injections as a weapon was 42-year-old Daniela Poggiali. She worked at the Umberto I Hospital in Lugo, where she disposed of more than 38 patients over a period of three months. Her victims were chosen if that person became too demanding and especially if they had relatives who annoyed her with their requests. When other nurses simply complained, Poggiali would say "leave them to me – I'll quieten them." And she did, with a fatal injection of potassium chloride. She was not a person to cross, as some of her nursing colleagues discovered. When they upset her, she would give their patients excessive doses of laxatives the evening before she went off duty so the nurses would have to deal with the mess the following day.

How many did she kill? In fact, Poggiali was on duty when 93 patients died over the two years she was employed at the hospital. This was twice as many as any other nurse, and it is now known that three of her victims died on a single day. What appalled many people were the photos she took of herself grinning next to their bodies.

Ms Poggiali had been a healthcare worker for 17 years. During that time, she had never shown signs of mental illness and relatively few of the traits normally associated with a psychopath. Such individuals may be devious but they can be charming and convincing, and while Poggiali came across as a rather cold person when dealing with her colleagues, she often flirted with the male doctors when she needed a favour. Eventually, the other nurses at the hospital noticed the unusually large number of patient deaths which occurred when Poggiali was on duty, and she began to be somewhat careless in her remarks. On one occasion, she even told a doctor who was treating a patient near the end of life that "two vials of potassium and his problem would be resolved."

Because she always got away with her killings, Poggiali began to do them more frequently. On 31 March 2014, Oriana Cricca was the target. She needed attention because her nasal feeding tube was leaking, but the nurse on duty was attending to another patient and so asked Poggiali to see to it. A few minutes later, Oriana died in agony. Two more patients who were being looked after by Poggiali died on 4 and 5 April. Although doctors performed autopsies, they found nothing unusual. But they too were beginning to suspect something was not right and Poggiali was moved from night shift duties to daytime ones. Even so, she continued her murdering. It would appear she now thought she was so clever that she could get away with it whenever she wanted. It was to be her undoing.

Some nurses suspected Poggiali was guilty of stealing from those she murdered, and not only the toiletries she found in their rooms but money as well. In some cases, as much as 100 euros went missing from a patient's purse or handbag. She also stole from the hospital itself and 12 such incidents were investigated, although without any charges being made. But who would suspect hard-working Poggiali? The doctors found her to be an excellent nurse, willingly volunteering to help, and who did so very competently. Her fellow nurses were not so easily fooled but, while they talked amongst themselves about the "Angel of Death," they did not report their suspicions to the authorities.

Poggiali's reign as a serial murderer came to an end in April 2014 with the unexpected death of 78-year-old Rosa Calderoni, who was being visited by her daughter Manuela Alci. When Poggiali arrived, apparently to attend to the old lady, she asked Manuela to leave the room while she did what she had to do. Ten minutes later, Poggiali said she had finished and Rosa was allowed back, to find her mother had a hypodermic syringe attached to her arm which was almost empty. Her mother's arm started to twitch erratically and violently, and she started rolling her eyes. Minutes later, she was dead.

This was not a death that could be easily passed off as something that was likely to happen to a woman of her age; Rosa was an elderly relative of Mauro Taglioni, the Director of Nursing at the hospital. A full investigation into the cause of death was undertaken but an empty ampoule of potassium chloride pointed immediately to the reason. Poggiali was later to say that she was not on good terms with Director Taglioni, but whether she was or not, she had committed one murder too many. She was arrested, and it later transpired that two vials of potassium chloride solution were unaccounted for in the hospital stores that day.

A doctor now withdrew aqueous humour from the dead woman's eye and tested it for potassium. The level was so high it indicated that she had been poisoned with this chemical to the extent that it would have caused her heart to stop beating. The doctors also found other evidence that Poggiali had been using potassium and they called the police. She was arrested and her iPhone confiscated. It was the images on that which provided the most convincing evidence of her behaviour, and they came in the form of pictures of her grinning over the bodies of her victims, in one of which she was giving the victory thumbs up sign. A colleague who took the photos later said that Poggiali was "euphoric" at the time, and that she had been browbeaten into taking them because she felt threatened by Poggiali, whom she knew could be very vindictive.

Poggiali is currently awaiting trial and no doubt her lawyer will claim that the deaths she was associated with can be explained in some other way. Meanwhile, the prison authorities say that she has had letters from admirers, including proposals of marriage.

Another nurse also awaiting trial is 50-year-old Vera Maresova. She has admitted killing six of her patients in the years 2010 to 2014 – five women and one man – who were in the intensive care unit at the hospital in Ramburk, a town of 11 000 inhabitants, in the Czech Republic. She was arrested after a 70-year-old woman died and was shown to have been poisoned with potassium. Her reason for the killings was to lighten her work load.

6.7 FAMILY MURDER, YOKOHAMA, JAPAN, 1989

Aum Shinrikyo, a religious cult, was founded by Shoko Asahara in 1984 and its beliefs are based on the Armageddon themes to be found in Hinduism, Buddhism, Yoga, and the Book of Revelation in the *Bible*. Cult members believe that we are living in the end of days. Rather oddly, it attracted a following among young intellectuals and other wealthy individuals. However, it was under attack by one particular lawyer, 33-year-old Tsutsumi Sakamoto, working on a class-action lawsuit being brought by families of those who had joined the cult, claiming that they had been brainwashed and duped into handing over their savings. He set up the Aum Shinrikyo Victims' Association.

Sakamoto had even challenged the cult leader to submit to blood tests to reveal the special power that he claimed only his body possessed. A blood sample was duly tested and revealed nothing special. Sakamoto informed the Tokyo Broadcasting System (TBS) about this and gave them an interview for a programme which he thought would be exposing the cult and its leader. TBS then approached Aum Shinrikyo for their opinion and they showed the cult leaders what Sakamoto had said about them. The leaders knew they had to act; enough was enough. Sakamoto had to be eliminated and in a way that would not raise any suspicions. It had to look as if he and his family had simply moved away from Yokohama and that he had lost interest in Aum Shinrikyo.

The task the assassins were given was to remove all trace of the lawyer, his family, and their belongings and, in so doing, make it appear to the outside world that they had simply gone away, leaving no forwarding address nor way of contacting them. And, just in case one of the bodies was eventually discovered, there had to be no tell-tale evidence of who they were nor how they had died. Injections of potassium chloride were to be the chosen agent. The plan worked, and for six years it appeared that the lawyer, his wife, and their young son had simply decided to pack up and leave Yokohama. That was in 1989.

I ought to remind the reader that the Aum Shinrikyo cult eventually came to be regarded as a terrorist organisation when it manufactured the nerve gas sarin and used it to attack commuters on the Tokyo underground system on 19 April 1995, where 12 people died and thousands were injured. The cult had done a trial run for this attack the previous June at a place called Matsumoto, where they released the deadly vapour at various sites and killed seven people and injured 200.

Back in 1989, their killing was on a more modest scale. At 3 am on the morning of 4 November, four men broke into the Sakamoto home without using force – they later claimed that a side door had been left unlocked – and they overcame him, his wife, and their 14-month-old baby boy and injected them with lethal doses of potassium chloride. The murderers were scientist Hideo Murai, martial arts master Satoro Hashimoto, along with Tomomasa Nakagawa and Kazuaki Okazaki.

The murderers then smashed the parents' teeth, so that they could not be identified by their dental records even if their bodies were found. They put their victims into three metal drums and dumped them in various rural areas, miles from one another, where they remained undiscovered for several years. Their mission was successful and no further class actions against the cult were initiated in the years following the disappearance of the Sakamoto family. In fact, their murders were only discovered when police were investigating the large scale attack by the cult on commuters on the Tokyo underground system. One of the perpetrators, Kazuaki Okazaki, confessed and at his trial and pleaded guilty to the three Sakamoto murders. Another of the assassins, scientist Hideo Murai, was himself murdered while in police custody.[††] Satoro Hashimoto and Tomomasa Nakagawa were convicted of murder in 2000 and sentenced to death and are still waiting for

[††]In 1995, Murai was stabbed by a man called Hiroyuki Jo who was a member of an organised crime gang. This happened while Murai was in police custody and while talking to reporters. He died on the way to hospital.

the sentence to be carried out. While waiting execution, Nakagawa has published articles renouncing his beliefs in the cult and asking forgiveness from the Sakamoto family for his crimes.

Today, Aum Shinrikyo as such no longer exists. In 2007, a group of its members split away to form Hikari no Wa (Circle of Light), while the original group re-named itself Aleph.

6.8 EUTHANASIA MURDER, WINCHESTER, ENGLAND, 1991

Lillian Boyes was a 70-year-old widow and had suffered from rheumatoid arthritis for more than 12 years. She had been in and out of hospital many times as the doctors tried to treat her condition but to little effect. Although her doctor at the hospital, Nigel Cox, had prescribed various treatments, her pain had become so bad that, in 1991, she was again admitted to the Royal Hampshire County Hospital in Winchester, but there was nothing they could do for her. It was said later that there was an unusually strong bond of affection and respect between Dr Cox and his patient. She could not praise him highly enough, while he greatly admired the courage of a woman who displayed remarkable cheerfulness and resilience over the years.

In the last few days of Mrs Boyes's life, events took a particularly cruel turn. As her pain intensified, she was given massive doses of diamorphine, the strongest analgesic in the medical armoury. But, highly unusually, it had little effect. She now wanted to die and so she refused all medical treatment. Indeed, the hospital chaplain, the Rev Richard Clarke, described her arms as no thicker than two of his fingers, and when the nurses moved her, the noise from her bone joints grinding together was said to be terrible to hear. She lived in constant intense pain such that any movement caused her to scream out.

Lillian pleaded with Dr Cox to put her out of her misery and it was evident that she had not long to live. On 16 August 1991, he injected her with a lethal dose of potassium chloride from two ampoules and she quickly died. After she had passed away, one of her sons who was visiting her at the time, thanked the doctor for what he had done. Cox entered details of the potassium chloride injection in the hospital log and gave the cause of death as bronchial pneumonia. That should have been the end of the matter, but the log was read by the rheumatology ward-sister who reported it to her superiors; the doctor had in fact murdered his patient and the hospital authorities decided to act and informed the police. Dr Cox was arrested.

His trial at Winchester Crown Court took place in September 1992 and lasted 10 days. The defence called several witnesses, of whom two leading rheumatologists told the court that Mrs Boyes was the worst case of rheumatoid arthritis they had ever encountered. They also said that she suffered from internal bleeding, septicaemia, and vasculitis, which is inflammation of the blood vessels.

The defence witnesses were asked what options were open to Dr Cox when the diamorphine failed to provide relief. One of them, Professor David Blake, appeared to find the question highly distressing and finally replied that he could not answer it. The other, Dr Alan Dixon, said he hoped he would have had the courage to do the same as Dr Cox. He added that the current legal position placed doctors in a curious predicament: they were permitted to administer drugs to ease suffering even if death was thereby hastened, but not to hasten death in order to ease suffering. It was, he said, "a razor's edge."

On the day she was given the potassium chloride injection, Mrs Boyes had but a few hours to live in any case. She had told Dr Cox that she wanted to die and her two sons, Patrick and John, who often sat with her, supported her decision. Dr Cox, who had previously refused to help her to die also promised her that she would not suffer, and on that particular day he felt that the time had come to end her agony by ending her life, just as she wished.

There was much discussion during the court proceedings of the ethics of medical intervention to shorten the life of terminally ill patients. Neil Butterfield QC, the leading Crown counsel, said that while it was a doctor's duty to minimise suffering, it was unacceptable both to the law and the medical profession to kill a patient to end that suffering. That debate still rumbles on in the UK.

Unfortunately, Dr Cox was found guilty of causing the death of Lillian Boyes and received a suspended sentence. His case was then reviewed by the General Medical Council in London and they decided that no further action need be taken other than a reprimand. He returned to his former job at the hospital in February 1993.

CHAPTER 7

Tetramethylenedisulfotetramine, a Mouthful Best Avoided

A word in **bold** *indicates that further information can be found in the Glossary. Only the first time the word appears in a chapter will it be so indicated.*

Tetramethylenedisulfotetramine, aka tetramine, TMDT, TETS, 4-2-4, and hereafter tetramine, was used as a rodent poison for several years before being banned worldwide in 1984 on account of its toxicity and potential for being misused. Nevertheless, it is still produced illegally because it is so effective and it has been responsible for several murders, especially in China, where it is known as Dushuqiang, which simply means "very strong rat poison." The number of cases of tetramine poisoning reported in China in the 20 years from 1991 to 2011 amounted to more than 14 000, of whom 932 people died, many of them children. Most of these cases were accidental poisonings but some were deliberate, such as that committed by a head teacher in Hebei Province who persuaded two young girls from a rival school to eat a tetramine-laced yoghurt, after which they quickly died. In Yunnan Province, there was an even larger attempt at poisoning infant children; again they attended a rival kindergarten, as you will read below.

More Molecules of Murder
By John Emsley

Published by the Royal Society of Chemistry, www.rsc.org

7.1 TETRAMINE

The discovery of the terrible potency of tetramine came as a result of sulfamide and formaldehyde being used together as a fire retardant on a rayon fabric. This was supplied to factories and it began to affect some workers quite badly. The new kind of apparently-safer rayon was manufactured by the chemical company Bayer in Germany, but it was workers in factories to which the new fabric was delivered who began to suffer headaches, nausea, vomiting and foaming at the mouth, while some even experienced seizures. Investigation soon revealed the cause: the two chemicals in the rayon had reacted with each other to produce a chemical which Bayer chemists identified as tetramine. Indeed, so powerful was it as a poison that Bayer patented it and began to market it as pesticide in 1953.

Tetramine was tested in the USA as a possible seed protectant, especially for tree seeds that were sown to reforest areas that had been denuded by felling or by forest fires. The new pesticide was not just to eliminate rodents that might eat the seeds, but to warn them off. The seeds were given a distinctive colouring that also tasted awful so that the rats would quickly learn not to eat them.

Just as the developers had hoped, more of the treated seeds germinated and survived to grow into trees compared to seeds treated with other pesticides, and it looked as if tetramine was to be the way forward. However, once the tree seedlings started to grow, they were also a source of food for other animals such as hares and these were dying as a result of eating the leaves. Not only that, but the young trees continued to be toxic for as long as four years. Clearly this new pesticide was too potent and it never became commercially available in the USA. That decision not to allow tetramine into the USA probably saved hundreds of lives, and not only of hares but of humans.

Tetramine comes as an odourless, tasteless white powder which is moderately soluble in water to the extent of 25 grams per litre. It was first made by H. Henecka and G. Hecht in 1949 by reacting sulfamide ($H_2NSO_2NH_2$) and formaldehyde (HCHO) in 60% sulfuric acid. Its toxicity was reported the following year, along with its **LD_{50}** for mice. It was eventually manufactured on a large scale for use as a pest controller, especially to kill rats, but it was a dangerous material even for the humans who made it and packed it, as was shown by the cases on record of such workers being poisoned by it. There were 12 cases of occupational exposure to breathing in tetramine by workers in a factory in China.

There was an accident at a factory in Hebei Province in China in July 1991 when 78 workers were accidentally poisoned. Although no one died, what was notable about the treatment of those exposed to the poison was that filtering their blood through activated charcoal could remove the tetramine from their blood – but this is only really effective when the doctors know what poison they are dealing with and often the symptoms were not diagnosed as being due to tetramine poisoning until it was too late and the victims were beyond help. Now, if tetramine poisoning is suspected, then forensic evidence can be obtained by **HPLC–MS** (high performance liquid chromatography–mass spectrometry).

The surprisingly lethal nature of tetramine continues to intrigue scientists, and many attempts have been made to understand its toxic behaviour and how this might be dealt with. A recent paper on the subject has shown that very little progress has been made in finding an antidote for the poison or a cure for those affected by it. The definitive paper is by Michael Shakarjian *et al.*, titled "Tetramethylenedisulfotetramine: pest control gone awry" and was published in the *Annals of the New York Academy of Science*, July 2016.

Tetramine is a neurotoxin. It binds to GABA receptors[†] in a way that means it cannot be removed by the body, and in so doing, it blocks the chloride channels which are part of the nervous system. This causes over-excitation of the central nervous system, which leads to convulsions and shows that tetramine is directly affecting the brain stem.

Radioactive carbon-14 investigations showed that most of the tetramine is removed from the body *via* the faeces and that takes about 72 hours, although some is even released as CO_2, showing that there are metabolic processes that can break the poison down to its smallest components. The half-life of tetramine in rats injected with the poison is 57 hours, whereas that given in their food had a half-life of 262 hours (11 days), which shows just

[†]Short for gamma-aminobutyric acid.

how difficult it is for the body's normal mechanisms for removing unwanted chemicals to deal with this toxin.

Depending on the amount of tetramine that has been taken, various components in the blood will be raised, such as the enzymes AST (aspartate aminotransferase) and ALT (alanine aminotransferase) which are released when the heart and liver are damaged. Tetramine is only slowly eliminated from the body and it can still be detected in the urine 10 days after being ingested.

7.2 TETRAMINE AS A POISON

Tetramine is on the World Health Organization's list of extremely hazardous pesticides, and it is considered to be even more lethal than sodium fluoroacetate. It is around a hundred times more deadly than cyanide. Tetramine is a highly effective rat poison. It has an LD_{50} of 0.20 mg kg^{-1} for mice, and for all mammals, including humans, the range is 0.1–0.3 mg kg^{-1}. This being so, then for the average 70 kg human, a dose of as little as 7 mg would likely prove fatal and a dose of 25 mg would certain kill. A dose as low as 6 mg has been reported to have caused death in a child.

The first symptoms of tetramine poisoning can appear within minutes of ingesting it, especially if it is drunk *via* water or milk. If eaten with food, there may be a delay of half an hour and, in one case, symptoms did not appear until 12 hours later. The mild symptoms are headache, dizziness, nausea, vomiting, and numbness around the mouth, and a general feeling of listlessness. The symptoms associated with a larger dose are fainting, seizures which may last a few minutes and keep on repeating, foaming at the mouth, and loss of bladder control. If the person goes into a coma then death may soon follow, and this will almost certainly be because of lack of air due to paralysis of the lungs.

There are no antidotes for tetramine poisoning; all you can do is treat the symptoms with large doses of benzodiazepines, which can boost the levels of the neurotransmitter GABA, and which will have a calming effect, and pyridoxine, which is vitamin B6. Animals that have been poisoned with tetramine have been treated with **dimercaptopropanesulfonate** (DMPS) and B6, and this will reduce convulsions and possibly prevent death.

Those who experience only mild symptoms of tetramine poisoning generally recover within a week. Those who take in a large dose will be dead within three hours. Severely affected people will suffer from liver enlargement and dysfunction.

Officially, the Chinese Government banned tetramine rat poison in 1991 and even warned those who still had some to hand it in to the authorities. This had only a lukewarm response, as shown by the number of tetramine poisoning cases that continued to happen. In the years following this attempt to regulate the poison there were 6000 people affected by it. It was so effective that it continued to be available illicitly. The problem with tetramine poisoning is that it is difficult to diagnose and, because it is officially banned, it is not something that medics suspect to be the cause of the convulsions which someone may be suffering from when admitted to hospital.

On 1 October 2003, the Chinese authorities threatened anyone storing or selling the poison with a prison sentence – even execution if it had been responsible for deaths. Many more people now handed their supplies in, but many others did not because it was so useful in controlling vermin. And, despite the warnings and less than a month later on 23 October 2003, a dozen or more customers of a barbecue stand in Shaanxi Province ended up in hospital when the owner of a rival barbecue stand poisoned his competitor's meat supplies with tetramine.

Just how dangerous tetramine can be was demonstrated by a 15-month-old girl who was crawling around the kitchen in her parent's New York apartment when she found some white crystals in the corner of the room. This was rat poison which her parents had brought from China. Within minutes the toddler was having seizures and writhing in agony. The parents rushed her to the A&E of the nearest hospital, explaining what had happened. The doctors there immediately began intensive therapy with a succession of drugs including lorazepam, which is given to treat seizures, phenobarbital, also to treat seizures and often given to epileptics, and pyridoxin, which is vitamin B6. They inserted a tube down her windpipe to help the child to breathe. This remained in place for three days, after which it was no longer needed, but by then it was apparent that the infant had suffered brain damage and would likely be blind. It was only months later that the real reason for the child's condition was revealed. The hospital sent samples taken from the child to the New York City Poison Control Center and they confirmed that the child had been poisoned with tetramine.

In 2004, a 53-year-old man went into a public library in Hong Kong, ate some food he had brought with him, and in which he had put some tetramine, then went into the toilets where he planned to die, as indicated by the suicide note he had in his pocket. There he collapsed, but was quickly discovered and whisked away to hospital where he was treated. Tetramine was detected in his urine, his blood, and the left-over food. Despite the best efforts of the medical staff, they could not save him and he died two days later.

For those admitted to hospital, the symptoms of tetramine poisoning, namely the convulsions they are experiencing, can be controlled to some extent by giving them an anticonvulsant such as benzodiazepine, ketamine, or barbiturates, while cleaning out the stomach with a gastric lavage. Detoxifying the blood of tetramine seemed to work for a while but then the level of the toxin in the blood would suddenly rise again, showing that the tetramine can infiltrate body tissues from which it can then be released.

Various treatments have been reported but none has been entirely successful, and the all-important antidote still waits to be discovered – if it will ever be found.

Tetramine is such a toxic agent that it could well be used as a terrorist weapon. It is easily made from relatively cheap and readily available chemicals. Once made, it is stable indefinitely and easily stored; it is colourless and odourless so is not easily detected; it dissolves readily in water and is tasteless; and it is deadly in tiny amounts. What no doubt deters would-be poisoners from using it is this last feature. A few grains of the poison on a person's fingers would be enough to kill were they to be transferred to food or into the mouth.

7.3 UNINTENDED MURDERS, CHINA, 2014

Nursery Wars: Two Dead, 76 Wounded

In the town of Shuanglongyin, in Yunnan Province in China, there are several nurseries for three- to five-year-old pre-school children, and most are privately run. The owners of the privately-run ones are always keen to attract children, and two of them, the Xiyangyang and the Jiajia kindergartens, were in direct competition and at loggerheads.

One evening, in March 2014, He Feng, the owner of the Xiyangyang nursery, broke into the Jiajia building and added tetramine to the drinking-water supply. The following morning, the cook at the school saw that there had been a break-in but, as nothing appeared to have been taken, she did not report it to the police.

Later that morning, some of the children started vomiting and one of the boys passed out while foaming at the mouth. The school rang the emergency services, and by the time they arrived, there were six more children in a critical condition and a full-scale emergency was declared. The children who were most seriously ill were rushed to the Qiubei People's Hospital. In the next few days, 76 children were detained there, two of whom died: four-year-old Yang Ziyi and five-year-old Zhou Yulin.

At first, it was thought that the children had been poisoned with sweets that had been bought as a treat because many of those who ate them were the ones who were ill. Tests on the sweet wrappers later showed they were not to blame. Eventually, it was deduced that the poison was in the water the children had drunk, and clearly this had been deliberately added by whoever had broken into the school the night before. He Feng was arrested on account of the threatening remarks he had recently made in a confrontation with the head of the Jiajia nursery.

So far, no one has been brought to trial for the children's murders.

7.4 MASS MURDERS, CHINA, 2000

Pastry Wars: 42 Dead, 300 Injured

Chen Zhengping ran a snack bar, the Zhengwu Pastry Bar, in Tangshan, in the Jiangning area of Nanjing city in the eastern coastal Jiangsu Province of China. Nearby was another snack bar, which his cousin owned, and this did much better business. Its patrons were the children of a local boarding school and the soldiers from a nearby barracks who enjoyed the various pastries which it sold, and especially its fried dough sticks, sticky rice balls, and sesame cakes. But its success was not to last.

On the morning of Friday 14 September 2001, the patrons of the snack bar were queuing to buy their favourite snacks and eating them as they went on to school or returned to the barracks. Soon they began to feel very ill, vomiting and bleeding from their mouths and ears in some cases. Some of the customers who ate their snack in the shop collapsed almost as soon as they had finished eating or just after leaving the premises.

It was soon clear that the local hospital had a mass poisoning on their hands as the children began to die, and sadly the staff were unable to save the lives of 38 of them. Ten soldiers also died that day. Investigation soon revealed that the victims had suffered from tetramine poisoning. Zhengping realised he was in trouble and fled, but a massive police response to the crime was ordered by the Chinese Communist Party and he was arrested two days later in Zhengzhou, 370 miles from Nanjing in Central China.

At his trial in 2001, the court heard that Zhengping had a criminal record, having been jailed for two and a half years for theft back in 1992. This time, he had entered his cousin's snack bar on the evening of 13 September at about 11 pm and added the tetramine rat poison to the flour and rice that was to be used the following day. The result was far in excess of what he intended, which was that people would simply be made ill by the food his cousin served and vow not to eat at the snack bar again, ideally transferring their custom to his bar. Of course, with a poison as deadly as tetramine he massively overdosed the flour and rice.

Zhengping was found guilty and sentenced to death and executed on Monday October 14 2002.

And still tetramine murders continue to take place. Another mass poisoning occurred at a funeral wake on 21 October 2003 in the mountainous region of Hubei Province in central China. A woman, Chen Xiaomei, who had been estranged from her husband and their son for many years, turned up at his funeral and added tetramine to the traditional funeral meal that guests ate. Ten of them died, including some of her own family members.

Even more recently, there have been cases of tetramine poisoning. In June 2015, it was reported that a 12-year-old boy had added tetramine to some cola, which he gave to his two sisters, aged eight and 14 at a village in Hunan Province in China. The two girls started vomiting and collapsed as they were walking home from school. By the time paramedics were able to attend to them, they were bleeding from their mouths, and by the time they reached hospital, they had stopped breathing. Their brother admitted to what he had done.

Also in the summer of 2015, the Xiao family who lived in Heze, Shandong Province in China lost their six-year-old son Li to tetramine. He was playing hide-and-seek with his friends and found some candy in the area at the rear of their house and ate it. No doubt it had been placed there to poison rats. Within minutes, he was screaming in pain and his parents rushed to his aid only to find he was bleeding from his eyes, ears, nose, and mouth. Although they rushed him to their local Juye County Hospital, he died later that day. His body was later shown to contain tetramine.

Part II
Natural Toxins

CHAPTER 8

Gelsemine and Cat Meat Stew

A word in **bold** *indicates that further information can be found in the Glossary. Only the first time the word appears in a chapter will it be so indicated.*

Yellow jasmine is often to be seen in the southern states of America and parts of southern China.† It is an evergreen vine which climbs round trees and can reach up to five metres in height. It has beautiful yellow flowers with funnel-shaped petals, and a pleasant fragrance, and it is the state flower of South Carolina. Its botanical name is *Gelsemium sempervirens* (Figure 8.1) and in China it is referred to as "heartbreak grass" because it has often been used to commit suicide. Indeed, its nectar can even be toxic to honeybees, which is perhaps not surprising because this plant produces a most deadly toxin, the molecule **gelsemine**.

The plant was classified by the famous botanist Carl Linnaeus in 1753 but it had been described in print as early as 1640 by John Parkinson, and although he tried growing it in his garden, it failed to flower. It is very much a plant of warmer climes, and even the plants grown at the famous Kew Gardens in London have never flowered.

---

†The yellow jasmine of Madeira is not the same plant but looks very similar.

More Molecules of Murder
By John Emsley

Published by the Royal Society of Chemistry, www.rsc.org

Figure 8.1 *Gelsemium sempervirens* is a beautiful yellow flower with funnel-shaped petals and is also known as yellow jasmine. © Manfred Ruckszio/Shutterstock.

Gelsemine was the poison which Huang Guang used to kill the Chinese millionaire Long Liyuan. Guang planned it so that Liyuan's death would be blamed on the restaurant where they had dined. What is rather strange to Western eyes is that the food he chose to poison was cat stew, of which Liyuan was particularly fond.

And what can kill a millionaire businessman in China, can also kill a millionaire Russian investor living in the stockbroker town of Weybridge in Surrey, England. But that death is most mysterious, as we shall see.

8.1 GELSEMINE

Gelsemine is present in the leaves, flowers, stem, roots, and rhizomes of yellow jasmine, and especially in the rhizomes. Rhizomes are the root-like growths which some plants produce and the means by which they spread. The rhizome sends up shoots to emerge as a new plant. As gardeners know, if you pull up a weed that has rhizomes, you also need to remove all of its rhizomes from the soil because each small segment left behind can create a new plant. Plants of this kind include ground elder, bindweed, and mint. Not that rhizomes are always to be discarded, because some spices such as ginger and turmeric come from them. What yellow jasmine offers in its 15–20 cm long rhizomes is just the opposite: a deadly poison.

This toxic ingredient was first isolated by Theo Wormley in 1870. It is one of the most lethal chemicals that Nature produces, a fact that has not been lost on some secret services for whom assassination by poisoning is part of their armoury. Its toxicity can be seen in its **LD_{50}** (lethal dose that will kill 50% of the test subjects), which for rats is 0.05 mg per kg of body weight, and for dogs it is 0.5. If human toxicity is similar to that for dogs, then for an average 70 kg human, a dose of 35 mg might well prove fatal.

A fatal dose of gelsemine will kill within an hour, although some victims have survived for as long as seven hours. If a person is to survive, then there needs to be rapid emptying of the stomach along with artificial respiration. How a victim of gelsemine poisoning reacts depends to some extent on the contents of the stomach, which may delay the onset of symptoms. Vomiting may eventually occur, but by then it may already be too late to expel the toxin, which will soon cause the victim to stop breathing and die. There is no known antidote as such for the poison, but if the signs and symptoms are dealt with, life can be saved.[‡]

Gelsemine works by strongly binding to, and activating, glycine receptors, which are present in parts of the nervous system. In general, this activation causes chloride ions to enter the nerve cells and this reduces the release of **acetylcholine** at the neural synapse. This results in lack of muscular stimulation, a condition known as flaccid paralysis. The symptoms displayed by its victims are giddiness (medically referred to as ataxia), droopy eyelids (ptosis), blurred and even double vision, and difficulty in breathing. These are experienced soon after imbibing the poison and often begin within minutes. Quite small doses can

[‡]**Atropine** acts as a kind of antidote and this is used to treat those poisoned with pesticides.

have a serious effect on breathing, causing this to become slower, while large doses can eventually cause it to cease altogether.

The symptoms reported by 58 people who were affected by gelsemine showed that all became dizzy, 86% had limb weakness, 67% vomited, 48% had difficulty breathing, and a similar number had blurred vision. Other symptoms, such as convulsions, stomach pains, diarrhoea, *etc.*, affected fewer individuals. Of another group of 315 people who had been affected by gelsemine, only 25 actually died. Most recovered within a week and some recovered within a day.

A case of accidental poisoning by gelsemine occurred in March 2007 when a retired couple living in Hong Kong gathered what they thought were leaves of the herb *Mussaenda pubescens*, which they took home to cook. They were unaware that they had gathered a lot of deadly jasmine leaves as well.[§] *Mussaenda* leaves are part of traditional Chinese medicine and said to control sweating caused by energy imbalances within the body. They are prescribed for colds, tonsillitis, and snake bites.

Together the couple ate about 60 grams of the leaves, but within 30 minutes of their meal, the 69-year-old woman became giddy, vomited, and passed out. Her husband rang for an ambulance to take her to hospital where she had a gastric lavage (stomach pump) and was put in intensive care for a few days until she recovered. Gelsemine was detected in her urine and stomach contents by means of **HPLC–MS** (high performance liquid chromatography–mass spectrometry). The husband, who was also 69 years old, was similarly affected but vomited up most of the herb broth within 30 minutes. He became dizzy and had difficulty breathing, but thankfully he did not lose consciousness and so was able to ring for help.

There were others, however, who researched gelsemine as a weapon of assassination. The Russian secret service has a special poison laboratory known as the Kamera, which was set up by Stalin in 1921. Its aim was to produce poisons that could incapacitate, or even kill, but in a way that would look like a natural illness with symptoms that would not raise suspicion of foul play among doctors and forensic investigators. So said Boris

[§]Their case is reported in detail by H. T. Fung *et al.*, *Hong Kong Journal of Emergency Medicine*, 2007, **14**, 221–224.

Volodarsky, a former Russian military intelligence officer. It is known that one of the poisons they produced was gelsemine, which killed quickly and left so little evidence in the body of the deceased that it was undetectable – but that was before the development of HPLC–MS.

8.2 THE MEDICAL USE OF GELSEMINE

Like so many plants, yellow jasmine has been used medicinally and was listed in pharmacopoeias around the world. It has been a part of Chinese herbal medicine for hundreds of years, to be used in the treatment of eczema, haemorrhoids, and even for leprosy. In the West, it was prescribed as gelsemium tincture, which was a solution in alcohol and water, and was given to treat various conditions such as tonsillitis, rheumatism, headaches, and even measles. It sometimes caused accidental death.

Gelsemine was extracted from the rhizomes of yellow jasmine by means of alcohol, and the solution so obtained contained around 0.4% of gelsemine. This was then evaporated to form a powder, which contained about 6% of gelsemine along with other molecules such as **scopoletin** and **palmitic acid**. A typical dose of the powdered root was 15–60 mg. When prescribed as a pure compound, the dose was 0.5–2 mg. It is still used as a traditional medicine and an updated account has been published by Dr Vandana Dutt and colleagues at the College of Pharmacy in Punjab, India.[¶]

"Gower's mixture" was the name given to the most common medical form of gelsemine, as prescribed to those suffering from migraine. It generally consisted of gelsemium tincture, sodium bromide, glyceryl trinitrate, and lemon syrup, and was taken three times a day after meals. It was seen mainly as a preventative and not to be given during an attack of migraine itself, when it would be of little benefit. Phenobarbitone was the more usual treatment when this happened.

Gelsemine came to the attention of medics in America in the 1840s when a Mississippi farmer gathered some yellow jasmine leaves and made herbal tea with them. As a result, he suffered the various symptoms associated with its toxic molecule. Despite

---

[¶]Their results were published in the journal *Pharmacognosy Review*, 2010, **4**, 185–194.

coming near to death, he recovered and noticed that the fever he had been suffering from had cleared up. A local doctor was intrigued by this and produced his own medicine from the plant, and this he called Electric Febrifuge.

In 1849, another doctor, Dr Porcher of South Carolina, informed the American Medical Association of the benefits to be obtained from *Gelsemium*, and by 1852, others were investigating it. A Dr E. M. Hale wrote a book called *Monograph on Gelsemium*, extolling its benefits. It was supposed to be particularly good in treating lung ailments such as asthma and whooping cough. Later, its main benefits were said to be for those suffering neuralgic pain, toothache, muscle spasms, rheumatism, and migraine headaches. In effect, gelsemium tincture was prescribed for several ailments where it might hide the pain associated with them.[‖] It was its ability to block a key messenger molecule which explained its use.

However, there is little scientific evidence of it being helpful in treating these conditions, and it is no longer prescribed medicinally, although a growth stimulant for pigs is based on gelsemine and said to improve the lean-to-fat ratio and increase the animals' immunity to other diseases. It is still possible to be treated homoeopathically with gelsemine and then it is recommended for flu, but such a treatment is, of course, perfectly safe since it contains none of the toxic agent.

When gelsemine tincture was prescribed, a typical dose would be 20 minims of the tincture, which in modern units would be 1.2 mL, and this would deliver around 1 mg of gelsemine. Such a dose would be unlikely to seriously affect a patient but its use did not go unquestioned. Indeed, gelsemine and its toxicity were described by Arthur Conan Doyle, the creator of Sherlock Holmes, in a letter to the *British Medical Journal* in its issue of 20 September 1879. Doyle was then a 20-year-old student at the University of Edinburgh Medical College.** He had read a report in the *BMJ* that someone had died having taken 75 minims of gelsemine tincture, while someone else had said that it was

‖It is also said to have anti-cancer properties against oesophagus cancer, and it is still being researched.

**He had also been writing his first detective short story, *The Mystery of Sasassa Valley*, which he submitted to a publisher on the 6th of September.

perfectly safe to consume 90 minims. Doses of this kind would deliver 4–5 mg of gelsemine.

Conan Doyle decided to test the efficacy of this new plant extract on himself, and in his letter he reported on the toxic effects of this medication which he had himself been using to treat his neuralgia. He had not suffered any of the side effects that others had warned about, so he had decided to increase the dose and report the effects it had on him.

He began his experiment on a Monday evening, and for that week he gave up smoking and drinking. On that occasion, he took a dose of 40 minims of the tincture, which would contain around 2 mg of gelsemine. This had no noticeable effect. The following evening he increased this to 60 minims (3 mg), again with no effect. On the Wednesday evening, he took 120 minims (6 mg) and within 20 minutes he was overcome with giddiness and his limbs felt weak, but he soon recovered. His pulse was weak but normal. On the Thursday, he repeated the dose of 120 minims, but this time, the giddiness was milder than the previous evening but he suffered blurred vision later that night. Then, on the Friday, he took 150 minims (7 mg) but it was not giddiness that was then the most noticeable; he suffered both from a painful headache and diarrhoea. On Saturday and Sunday he took 200 minims (10 mg). He then had persistent diarrhoea, severe headache, and feeble pulse. He decided that enough was enough and the experiment ended.

8.3 MURDER, BRITISH INDIA, 1913

Life in the far-flung British Empire was not without its interesting murders, one of which involved gelsemine, and it may have been this which gave Agatha Christie the idea for including it in one of her novels.

The true murder story began in 1909 in Agra, India, where Augusta Fullam lived with her husband Edward. The marriage was not a happy one and she had fallen in love with a Lieutenant Henry Clark, who worked in the medical section of the local military hospital. Henry was the son of a mixed-race marriage, and he was married to a lady called Louisa. The two lovers planned to dispose of their partners so they could be together and maybe even marry. This was revealed when the letters which passed between them came to light as result of the investigation into the death of Louisa, who had been murdered by intruders into her home.

The lover's plot began with Augusta trying to poison her husband with arsenic – which Henry provided – by putting it into the Sanatogen tonic wine of which he was fond. This made him ill several times but didn't have the desired effect. Clearly something stronger was needed, and so Henry sent her some gelsemine and that did the trick. When her husband died, Henry came round and signed the death certificate. So far so good.

Now it was Henry's turn to dispose of his spouse, which he did in a rather violent way, but for which he would have an unquestionable alibi. Henry paid a local hit man to come and murder his wife while he was at work. The man, with three of his friends, broke into the house and slaughtered Louisa with a sabre. Now the police were involved, although on the face of it, they were dealing simply with a burglary that had gone disastrously wrong.

However, the affair between Henry and Augusta had not gone unnoticed among the local residents, and this was confirmed when the police found letters in Henry's house which revealed what the lovers had been doing. They were arrested and put on trial. Both were found guilty and sentenced to death. Henry was hanged on 26 March 1913 but Augusta's life was spared because she was pregnant with Henry's child. She died in Naini prison the following year. The men who slaughtered Louisa were also executed.

Such was the notoriety of the above murders that books were written about the case, and no doubt one of these was read by Agatha Christie. She then used gelsemine as the poison in her whodunit novel *The Big Four*, which was published in 1927. The person murdered with gelsemine was a Mr Paytner, and as he was dying, he had written the words "yellow jasmine" across the newspaper he was holding and he wrote with his finger dipped in ink. This plant was growing in the garden of the house where he died, but the words meant nothing to those who found the body, except they revealed the cause of death to Christie's famous detective Hercule Poirot.

Other well-known poisons, like cyanide and **strychnine**, featured in the book but it was gelsemine that was new to the reading public. Christie even explained that it was extracted from yellow jasmine and she listed the various molecules which this plant produces. She even gave its chemical formula, albeit one nitrogen short, and described correctly the effects it had on the body's central nervous system.

8.4 MURDER, CHINA, 2011

Cat stew is not a culinary delight in the UK, nor ever likely to be, despite its supposed health benefits, but it is eaten in China and other countries out East. More than three million Chinese cats end up being eaten every year, although this might eventually come to an end if the Chinese Animal Protection Network has its way. That organisation campaigns against cat meat and has organised demonstrations outside restaurants which serve it. Maybe if they had protested outside a particular restaurant in Yangjiang in Guangdong Province, southern China, then they might have prevented a murder.

Cat meat is reputed to fortify the body. In Korea, it is often served as stew and thought not only to be a tonic, but a remedy for neuralgia and arthritis. However, China is the main country where cat meat is eaten. Cat stomach, intestines, and thighs are consumed often in the form of meatballs served with stew, and in Guangdong Province, cat meat is the main ingredient in the dish referred to as "Dragon, tiger, phoenix." This is prepared from a freshly-killed animal and it needs to be cooked for several hours to make the meat tender. This type of stew is probably the dish that three diners ordered one fateful day in 2011.

Yangjiang is a city of more than two million people and, on 23 December 2011, 49-year-old millionaire tycoon Long Liyuan went out for what he thought was a business lunch with Huang Guang and a colleague Hung Wen. Liyuan had dined at the restaurant previously and particularly liked their cat stew. Guang was deputy director of the Guangdong Agriculture Department and had collaborated with Liyuan, helping him to acquire a profitable forestry lease in 2009. As part of the deal, Liyuan lent money to Guang, amounting to 2.45 million yuan (about £27 000), thinking he too was investing in the region and that he was planning further schemes which might well benefit them both. However, Guang had gambled it all away and there was no way he could repay the loan, so he decided to murder the millionaire and clear the debt that way. He devised a way of doing this so that he would not be blamed for the death. Instead, it would be seen as due to food poisoning.

On that fateful December day, Liyuan sat down to lunch with Guang and Wen and they ordered the cat stew. Unbeknown to his fellow diners, Guang had with him some gelsemine powder which he had obtained from a local herbalist and he planned to put some in the cat stew at the restaurant. Clearly, he had to ensure that, of the three men who ate the stew, only Liyuan should imbibe a fatal dose, while Guang and Wen would consume just enough to indicate they too had been poisoned but not to anything like a lethal extent.

Guang's plan involved his going to the restaurant kitchen after they had ordered their cat stew with the excuse that his guests needed to be served quickly as their time was limited. No doubt he offered to take the stew to their table, and this gave him the opportunity to add a little gelsemine to two of the bowls of stew and a lot to the third bowl, which he then placed in front of Liyuan.

Soon after eating the stew, the three men began to complain of feeling ill and Liyuan showed such extreme symptoms that an ambulance was called for. They were taken to hospital for treatment and it was there that Liyuan died of cardiac arrest. The other two survived.

Liyuan's family were devastated by his sudden death. They suspected he had been poisoned and offered a reward of 100 000 yuan (about £1800) for information. Their chief suspect was Guang because they knew he was in debt to Liyuan. The police focussed on Guang and their enquiries soon revealed that, indeed, he had a strong motive for wishing Liyuan dead and he was arrested on 30 December.

At his trial in January 2012, Guang was found guilty and fined 100 000 yuan (about £1800) and ordered to pay compensation of 580 000 yuan (about £6300) to Liyuan's family. He was also sentenced to death, a punishment which Liyuan's family accepted as just.

8.5 MURDER, WEYBRIDGE, SURREY, 2010

Wealthy Russian oligarchs who come to live in the UK sometimes end up dying in rather suspicious circumstances. Boris

Berezovsky, aged 67, was found hanged in his bathroom at his Ascot home in March 2013. Arkady ("Badri") Patarkatsishvili, aged 52, who also lived in Surrey, died of a heart attack in 2008. He had told friends that there had been threats to murder him. Then it was to be the turn of multi-millionaire 44-year-old Alexander Perepilichnyy to die unexpectedly in 2012, and that's when gelsemine entered the picture.

Alexander Perepilichnyy had studied maths and physics at Moscow's Physical and Technological University and graduated in 1991, just after the fall of the Soviet Union. He got a job selling computers and was so successful that he soon bought himself a black Mercedes. Within a few years, he had become very rich by investing his savings on the stock exchange and eventually ended up a multi-millionaire.

Perhaps Perepilichnyy was too honest for his own good. Russia was undergoing a fundamental change from communist dictatorship to unbounded capitalism. He was appalled at the corrupt goings-on and it was common knowledge that there were Russian tax officials who had conspired with organised criminals to defraud the state of more than $200 million. Perepilichnyy knew of such activity at an investment fund called Hermitage Capital Management. Its accounts had been plundered and this had enabled prominent Government officials to salt away $11 million in a Swiss bank account. Perepilichnyy had informed the Swiss about this theft, despite being warned not to, and it is said that he was thereafter on a hit list.

One person who did very well out of all this financial scheming was Olga Stepanova, a senior tax official. Not only did she have a £3 million house built near Moscow but had bought two luxury villas in Dubai. This came to light after the arrest and death in custody of Sergei Magnitsky, a lawyer associated with the Heritage group and who was based in Moscow. He was about to reveal what was going on and he was arrested on a trumped up charge. He died later from an untreated inflammation of the pancreas, which was said to have been caused by a beating.

The Heritage fund itself was run by a US-born financier, Bill Browder, and Perepilichnyy told him what had been going on. He also informed the Swiss about the missing money. Browder decided to move the fund to London. He was aware that

Perepilichnyy was also at risk and was glad to see him leave Moscow.

Perepilichnyy came to the UK in 2010, along with his wife Nataliya, and their two children, and in November that year he bought a large house in the St Georges Hill gated estate in Weybridge. It is reputed to have cost £3 million. He was determined to keep a low profile in his new home, but he was aware that there were people back in Moscow who were very much aggrieved at what he had done to them. He now took out life insurance totalling more than £3 million with various insurance companies in order to protect his family. Before he could be insured for these amounts, he had undergone extensive medical tests and he was clearly a fit individual and with no underlying health problems so there was no reason not to insure him. He kept himself fit, and on most days the 44-year-old multimillionaire went jogging.

On 6 November 2012, Perepilichnyy went to Paris, ostensibly to meet up with an old friend, but clearly to meet other people as well. This appears to be borne out because he booked rooms at three luxury hotels. Whom he met in Paris has not been revealed, nor what his meetings were about, but he returned to London on 9 November.

The day after he returned to Weybridge, Perepilichnyy went jogging around the St George's estate but suddenly he collapsed at 5:15 pm outside his house and died 25 minutes later, and before paramedics arrived. It was said that he had a puncture mark on his neck. Had he been targeted by someone waiting for him and who knew of his daily routine of jogging? It seems likely that the person accosted him and jabbed him with a spring-loaded hypodermic needle delivering a fatal dose of gelsemine, which would quickly render him unconscious.

To those who were unaware of Perepilichnyy's background, it appeared that jogging had put too much of a strain on Perepilichnyy's heart. On the face of it, this was simply a case of a heart attack, which may have been unexpected but was a natural cause of death. The police became involved but did little other than note what had happened. However, there were those who wanted him dead, and there were those in the UK who knew he had been threatened and was very much at risk.

A post-mortem on Perepilichnyy's body was carried out by a Home Office pathologist but there was no indication that he had died other than by natural causes. Then, in 2014, toxicological samples from his body were submitted for analysis. A plant expert, Professor Monique Simmonds, from the Royal Botanic Gardens at Kew, south-west London, was asked by the company that had insured Perepilichnyy's life to analyse his blood to see if there was any suspicious toxin present that might account for his demise.

Simmonds found evidence that suggested the presence of gelsemine, which meant that he had been killed deliberately, and an inquest into Perepilichnyy's death was set for 18 May 2015. But how could this toxin possibly have got into his body? It is hardly likely that someone could have poisoned his food before he went jogging because that would soon make him feel too ill to undertake such exercise. Clearly there was more to Perepilichnyy's death than was originally realised and the senior coroner for Surrey, Richard Travers, postponed the inquest to 21 September 2015. Then, more information into Perepilichnyy's background came to light and this indicated that there were those who had wanted to see him dead. There were also those who wished to hear what the coroner had to say, such as the Hermitage Capital Management people who were now represented by a lawyer, Henrietta Hill. She said that documents that Perepilichnyy had sent to Hermitage contained "explosive" information but this was not to be revealed and the inquest was adjourned for a second time.

The re-convened inquest was held in January 2016 but it was said that forensic results were still outstanding. Now, the Hermitage Capital fund had employed Geoffrey Robertson QC to attend, and he revealed that Perepilichnyy had been afraid for his life and he had been providing the hedge fund with information. Robertson also indicated that Perepilichnyy might have been in contact with the security services, and indeed, subsequent events pointed in this direction. Robertson maintained that the Surrey police investigation of Perepilichnyy's death was a "cover-up," as he described it. He complained that he was unable to examine 45 documents that might have bearings on the case, and he said that the local police had been told not to record any interviews with those involved in the case, a procedure which the barrister said was illegal.

In fact, the police requested that those documents should be granted PII status (Public Interest Immunity) on account of their being a threat to national security. The fact that they have been allowed to classify them this way means that the information they contained might reveal things about the case that were better kept secret. But what could that be? Charlotte Ventham, acting for the Surrey Police, denied that the force was guilty of a cover-up and said they refuted these allegations.

Robertson likened what had happened to Perepilichnyy to the murder of another Russian whistle-blower, Alexander Litvinenko, who had been assassinated by Russian agents in 2007. They poisoned him with polonium-210, only a few millionths of a gram of which are enough to kill. However, the eventual identification of that toxic agent by British forensic scientists made it an unlikely poison for future use because its radioactivity reveals its presence.

The coroner postponed the inquest yet again and a date was fixed for 29 February 2016, but then it was postponed a third time without any evidence being heard. The inquest was set for September 2016 but the PII card was played again and it was moved to 13 March 2017. This would follow legal arguments about PII, which were heard at the High Court in November 2016 when an order was issued, requested by the Home Secretary Amber Rudd, to prevent the disclosure of sensitive material at the forthcoming inquest. This order covered 49 documents relating to the case.

At the resumed inquest, on 13th March 2017, which rather strangely did not involve a jury, there was evidence suggesting that Perepilichnyy had been poisoned *via* some sorrel soup, which he had eaten the day he died. This had been difficult to prove because his stomach contents had been 'flushed away' following the post-mortem, although there was evidence of gelsemine in some material extracted from his stomach lining. The inquest was again adjourned. A full hearing of the case is to be heard in London before judge Nicholas Hilliard QC in June 2017.

CHAPTER 9

Strychnine and Cream

A word in **bold** *indicates that further information can be found in the Glossary. Only the first time the word appears in a chapter will it be so indicated.*

Ask someone to name the three poisons most used to murder and they will probably say arsenic, cyanide, and **strychnine**. I dealt with arsenic in my book *Elements of Murder*, cyanide in *Molecules of Murder*, and now it's the turn of strychnine. Its reputation explains why the current chapter is the longest in this book.

Strychnine was once used to kill vermin and, because it was readily available, it was also used to kill humans. The chances of this happening today are remote because in many countries it is now illegal to sell strychnine, although it can still be bought in Pakistan and the USA. When it was legal to sell it in the UK, there was Marsden's Vermin and Insect Killer, which consisted of strychnine mixed with flour and coloured with a red dye. It was meant to be spread on bread and left on the kitchen floor at night to poison rats and mice, and these would sometimes be found dead nearby, so quick acting was the poison. The use of strychnine was finally outlawed in the Cruel Poisons Act of 1962.

Strychnine is a powerful poison causing painful convulsions and death within the hour. It distorts the spine backwards into

More Molecules of Murder
By John Emsley

Published by the Royal Society of Chemistry, www.rsc.org

an arch and causes the face to freeze in a grimace, and these deformities persist after death. What would-be murderers had to disguise was its bitter taste,[†] and this they could do by mixing it with something sweet tasting or by enclosing it in a capsule, to be swallowed like a tablet. The latter method was used in some famous murders, such as those committed by a serial poisoner of London prostitutes and by a supposedly helpful widow of Michigan, who despatched several family members with it.

In 2015, this poison was involved in a curious suicide by an unknown man on a lonely Lancashire moor near the site of an air disaster in 1949.

9.1 STRYCHNINE THE MOLECULE

Carl Linnaeus gave the name *Strychno* to a group of trees and climbing shrubs of which there are almost 200 species, and they grow in the warmer parts of the world. The seeds and bark contain strychnine. The best known is the *Strychnos nux-vomica* tree, which grows in the Malabar region of south India and in Sri Lanka. (*Nux* means nut. The *vomica* part of the name is derived from the Latin word for sores, which the juice of the plant was said to heal.) Its toxic nature had long been known, and it was made use of as a pest exterminator. In 1558, the Italian Giambattista della Porta described, in his book *Magia Naturalis* (Natural Magic[‡]), how filings from a *Strychnos nux-vomica* nut could be mixed with butter and fed to a dog, which would quickly die.

The tree grows to over 10 metres high and has a crooked trunk. Its greenish-white flowers produce orange-red fruits (Figure 9.1), each containing five disc-shaped seeds and these are a convenient source of the poison. In the 1880s, almost 500 tons a year of these were imported into London. Most were used to poison animals that were a threat to food supplies, such as birds, rats, and mice, or were a nuisance, such as dogs and cats.

Strychnine is an alkaloid and was first isolated in 1818 by two French chemists, Joseph Caventou and Pierre-Joseph Pelletier.

[†]Its bitterness can be detected in water at concentrations as low as 8 ppm (parts per million).

[‡]It also dealt with metallurgy, geology, magnetism, optics, medicines, and poisons. It ran to several editions and was translated into other languages, including English, in 1658.

Figure 9.1 *Strychnos nux-vomica* fruit containing the poison strychnine. © wasanajai/Shutterstock.

It was eventually analysed as $C_{21}H_{22}N_2O_2$, but its molecular structure was to remain unknown for the next 125 years. It was finally deduced in 1945 by Sir Robert Robinson[§] at Oxford University, and it was first synthesised in the Harvard laboratory of Robert Woodward[¶] in 1954, but he required 28 separate chemical reactions in order to construct the molecule and then it was obtained in only tiny yield. What still intrigues chemists is the ease with which *nux vomica* can make the molecule from the raw materials that Nature provides, namely water, carbon dioxide, ammonia, and sunlight.

Strychnine's molecular structure was confirmed as correct by X-ray crystal analysis in the 1950s. Today, it can be detected, identified, and measured by **HPLC–MS** (high performance liquid chromatography–mass spectrometry).

Strychnine is obtained by grinding the nuts of *nux-vomica*, which contain about 1% of the toxin, and extracting it by means of a solvent like alcohol. It is not very soluble in water; it requires 6 litres to dissolve only 1 gram. However, its salts are much more soluble, such as strychnine hydrochloride, which is soluble in water to the extent of 1 g in 40 mL, as is strychnine sulfate (1 g in 50 mL).

[§]He won the 1947 Nobel Prize in Chemistry.
[¶]For which he too won a Nobel Prize in 1965.

9.2 THE EFFECT OF STRYCHNINE IN THE HUMAN BODY

The human body recognises strychnine as a molecule it does not need and which it must dispose of rapidly. The liver immediately begins to do that and it has a half-life in the blood of only 50 minutes. However short a time this might appear, it is not short enough if a dose of 30 mg or more is consumed, and death will quickly occur.

Strychnine poisoning begins with muscle stiffness and cramps followed by convulsions and distortion of the limbs as it affects the spinal cord. This is responsible for the bridge-like arching of the back which is the most noticeable symptom, with the body resting on the back of the head and the heels; a posture so extraordinary that it was recorded in paintings as long ago as 1800. Another feature is *Risus sardonicus* (mocking smile), a facial grimace with raised eyebrows and extended lips caused by the contraction of the muscles, and this makes speaking difficult. The symptoms of strychnine poisoning were often confused with the grand mal seizures of epilepsy, or with tetanus.

Strychnine targets the central nervous system and its network of nerve fibres. The junctions between these are called synapses and it is between them that the messenger molecule **acetylcholine** is sent to pass the signal from one to another. Having done so, it then needs to be deactivated if it is not to keep repeating the signal, and this deactivation is done in part by the action of glycine. Strychnine blocks the glycine switch and so the nerve just keeps responding to the acetylcholine because there is nothing to turn it off. The spinal cord and brain stem are most affected and the result is spasms and convulsions.

Once it has entered the body, strychnine is rapidly absorbed and starts to act, and unless medical assistance is quickly provided, then there is little chance of survival. If 50–100 mg has been ingested, death is likely to come within the hour and not later than two hours.

Symptoms of strychnine poisoning appear within 15 to 30 minutes and start with a tingling sensation in the arms and legs, and this progresses into muscle spasms and convulsions of the whole body. After a while these might calm down, only to reappear after a few minutes as an even more severe attack begins. Strychnine decreases the pH of the blood below its normal value of 7.4, which is slightly alkaline, and if this falls below 6.8 then a person is likely not to survive. It also raises the body temperature to around 43 °C (109 °F).

The body becomes short of oxygen because of the demands being made by muscle contractions, and in any case, the muscles that control the action of the lungs are also malfunctioning.

There is no antidote for strychnine. All that can be done is to alleviate its symptoms while the kidneys work to remove the poison from the body, and this takes time. Muscle relaxants such as **diazepam** (aka Valium) are used to calm the central nervous system to suppress convulsions and relax muscles. The victim needs to be kept in a quiet, dark room to prevent the brain from generating signals which pass along the central nervous system.

Giving the patient a drink containing activated charcoal can help absorb the strychnine in the stomach and so isolate it, provided it can be given within 20 minutes of the strychnine being ingested. Activated charcoal is a form of carbon that has large cavities into which other molecules can penetrate and be held fast. (It is the active component of gas masks.) Its effectiveness was demonstrated as long ago as 1831, when a Montpellier pharmacist called Pierre Tourey consumed almost a gram of strychnine (more than 10 times the fatal dose) mixed with 15 g of activated charcoal and suffered no ill effects. The stomach should also be washed with an oxidising agent such as potassium permanganate to destroy the strychnine or with a precipitating agent such as tannic acid, with which strychnine forms an insoluble salt.

If a person can survive being poisoned with strychnine, then there is usually no lasting physical damage. Almost all strychnine will be removed *via* the urine within 24 hours.

So how toxic is strychnine? The **LD_{50}** for dogs is 0.5 mg per kg, and if this value is assumed to be the same for humans, then 35 mg should have a 50 : 50 chance of killing a typical 70 kg person. A dose of twice this amount would almost certainly kill. (Rats appear to be able to cope with this poison much better than dogs because for them the LD_{50} is 16 mg per kg.)

9.3 STRYCHNINE AS A TONIC

Strychnine may be deadly but this did not exclude its being prescribed medically or taken as an ingredient in over-the-counter health tonics. For many years, small doses of strychnine were approved by doctors under the impression that it acted as a stimulant to the nervous system. Indeed, in Victorian times, strychnine was said to be useful in the treatment of deafness, headaches, intestinal worms, rheumatism, diabetes, hernia, and cholera. This upsurge of interest came about as a result of books written in the 1830s which extolled the virtues of this new drug, and in 1836, it entered the British Pharmacopoeia (BP). As late as 1934, the BP still said that strychnine increased appetite and improved muscle tone, leading to benefits for the heart and lungs. It was also advocated for chronic constipation and could be given after surgery in the form of an injection. Adolf Hitler was prescribed strychnine in the form of Dr Koester's Anti-Gas Pills as a remedy for the indigestion and the excessive farting to which he was prone.

The supposed tonic action of strychnine was justified in a letter published in the *British Medical Journal* edition of 11 March 1944 and written by a doctor, W. F. Anderson MD. In this, he claimed it had benefits for the gastric system, which he said he had researched. He claimed that strychnine increased the production of digestive juices and speeded the emptying of the stomach.

It was still listed in the BP of 1959, and the recommended dose was 2–8 mg to counteract muscular fatigue. Some sports people still use it, which is why it is still included in the list of drugs to be tested for at certain events. In 2001, a weightlifter in India was banned because he had used it. At one such contest, more than half the weightlifters withdrew once they realised that they would be tested for this drug.

As long ago as the 1890s, strychnine started to be taken by athletes because its convulsing symptoms were thought to tone muscles and so enhance performance in the field and on the race track. In the 1904 Olympics, held in St Louis, USA, the marathon was won by Thomas Hicks.[‖] He made use of a tonic of raw egg, brandy, and a few crystals of strychnine sulphate.

[‖] He was not the first runner to enter the stadium. That had been Fred Lors of New York City but he had been helped to complete the run by a spectator who gave him a lift in a car.

However, he had to be helped over the finishing line because he collapsed just before he reached it.

Over-the-counter tonics once contained strychnine, which many took in the belief that it restored vitality to the body. One particular tonic was Fellow's Compound Syrup of Hypophosphites, which cost seven shillings for a bottle of 15 fluid ounces (420 mL). The price equivalent today would be £38. The recommended dose would provide 1/64 grain (1 mg) of strychnine. The most popular of such tonics was Metadone, produced by the US Company Parke Davis, and although this tonic is still available, it no longer contains strychnine.**

The reference work on drugs known as Martindale noted that the drug was readily absorbed from the stomach into the bloodstream and from there it passed into various tissues within five minutes. It warned of the symptoms that it produced, with special emphasis on the contraction of the muscles of the diaphragm which eventually caused breathing to stop. Martindale cited a case in 1971 when a 20-month-old child had eaten some laxative tablets that contained strychnine nitrate and had ingested about 20 mg of the poison. The child's life was saved by intravenous injections of diazepam. Eventually, strychnine was deduced to have no health benefits at all, and by 1971, it was no longer prescribed by doctors.

9.4 STRYCHNINE IN POPULAR CULTURE

Strychnine got its reputation partly through its role in famous novels, beginning with Alexander Dumas' *The Count of Monte Cristo*, published in 1844. In this, he reports deaths caused by strychnine poisoning. Arthur Conan Doyle's indomitable detective Sherlock Holmes encounters strychnine in *The Sign of Four*, published in 1890. His side-kick Dr Watson confirms that it is responsible for the death of Bartholomew Sholto, and this he deduces from the unusual grimace on the dead man's face.

Not every author used it as an agent of death. In fact, it received approval from H. G. Wells in his 1897 best-seller,

Today it consists of vitamin B_1 and the essential metal elements calcium, potassium, sodium, and magnesium in the form of their **glycerophosphate salts.

The Invisible Man. The title character Dr Griffin found it to be beneficial:

> Griffin had a little breakdown. He started to have nightmares and was no longer interested in his work. But he took some strychnine and felt energized.

Strychnine murders occur in Agatha Christie's whodunits, as in *The Mysterious Affair at Styles*, published in 1920, in which a Mrs Emily Inglethorp is poisoned with it. Ten years later, it featured again in her book *The Coming of Mr Quin*, and then in *How Does Your Garden Grow?* (1935). James Herriott had this poison in his novel *All Creatures Great and Small*, written in 1972 and made into a popular TV series. The most recent novel in which strychnine appears is *Mr Mercedes* by Stephen King, published in 2016.

Strychnine's toxic reputation even featured in some popular songs. It occurs in Tom Lehrer's "Poisoning Pigeons in the Park," released in the 1960s, which includes the verse:

> My pulse will be quickenin'
> With each drop of strychnine
> We feed to a pigeon.
> It just takes a smidgin
> To poison a pigeon in the park.

While the Welsh glam-rock band Manic Street Preachers released the song "You Love Us" in 1991 and it included the lines:

> We won't die of devotion.
> Understand we can never belong.
> Throw some acid on the Mona-Lisa's face,
> Pollute your mineral water with a strychnine taste.

Strychnine played a part in two of Alfred Hitchcock's films: *A Blueprint for Murder* (1953) and his most famous film of all, *Psycho* (1960), in which we finally discover that the psychopath Norman Bates has poisoned his mother and her lover with it. Wes Anderson's 2014 film *The Grand Budapest Hotel* includes a murder with strychnine, showing that this poison has still not lost it nefarious reputation.

9.5 MURDER, LONDON, 1830

The current Poet Laureate, Andrew Motion, has written a biographical novel about Thomas Griffiths Wainewright called *Wainewright the Poisoner: The True Confession of a Charming and Ingenious Criminal*. Like others before him, Motion was intrigued by a man who came from a privileged background, who had above average talent as a writer and painter, and yet who resorted to forgery and murder. Motion throws doubt on his being a serial killer and suggests he had an accomplice, namely his wife Eliza.

Wainewright was born in Richmond, Surrey, on September 1794 and died in Hobart, Tasmania, in August 1847. He was an acknowledged artist, famed for his portraits, some of which still grace the walls of homes and galleries. He was in Tasmania because he'd been found guilty of forging cheques for large amounts in order to finance his extravagant lifestyle, and he was sentenced to transportation to the colonies. He had also resorted to poisoning to inherit money, and for this he used strychnine.

Wainewright's mother had died soon after giving birth to him, and his father died when he was only a child. He came from a wealthy family of lawyers, and he was brought up first by his grandfather, at his home in Turnham Green, London, and then by his uncle, George Griffiths. He was educated at Greenwich Academy, from which he entered the army. On 13 November 1817, he married Eliza Frances Ward, and he put into a trust fund the dowry given him by her parents. He eventually realised that an army life was not for him, and in 1819, he resigned his commission. He returned to London where he became a member of a circle of writers and artists that included William Wordsworth and Charles Lamb.

To begin with, Wainewright wrote articles for magazines and especially for *The London Magazine*, and these were much admired. He was financially independent in any case, having inherited £5200 from his grandfather (worth more than £1.5 million today) and this brought him an income of £200 a year (= £100 000) which would have been enough to lead a comfortable middle class existence, but it was not enough to fund the life he was leading.

Wainewright also aspired to become an artist and trained as such under John Linnell. His paintings were judged of sufficient quality to be exhibited at the Royal Academy. He also illustrated books. But Wainewright and his wife spent lavishly, and to continue this way of life, he withdrew large amounts from the trust fund by forging the signatures of its four trustees. By 1828 that money too was gone and the couple were virtually bankrupt. They moved to live with ailing Uncle George, who died within a short time and who left everything to Thomas. That inheritance too was soon spent. In 1830, Eliza's mother made a will in her daughter's favour and she died a week later. It seems more than likely that Wainewright poisoned them. Indeed, he confessed on his deathbed to poisoning his mother-in-law, and he said he did it because she had ugly ankles.

Wainewright then insured 20-year-old Helen, a step-daughter of Eliza's, for £18 000 (equivalent to more than £5 million today) and she died a few months later. The insurance company refused to pay on the grounds that there were errors in the policy. Wainewright took them to court – and lost the case. He was by now suspected of forgery and so deemed it expedient to go and live in France until things quietened down. While there, he insured the life of the father of a girl with whom he began a relationship and then murdered him, thereby collecting £3000 insurance money (£800 000 today). This was soon spent and early in 1837 he returned to London but was arrested and charged with the forgery of the power of attorney documents. He was found guilty and sentenced to transportation to Tasmania. He travelled on the convict ship *Susan* and arrived there in November 1837.

For a while, Wainewright worked on a road gang and then as an orderly in a hospital. His skill as an artist was eventually recognised, and he was able to find work painting portraits of the local gentry and he produced more than a hundred of them. He was given a conditional pardon by the Governor in November 1846 and died the following August, aged 50, of what was described as apoplexy. Later that century, he was seen by some as a tragic figure – Charles Dickens and Oscar Wilde wrote about him – although it was always recognised that he had poisoned people with strychnine, which he had had in his possession when he was arrested in London.

9.6 MURDER, LAMBETH, LONDON, 1891

Serial poisoner Thomas Neill Cream was born in Glasgow in May 1850, the eldest of eight children. The family moved to Canada when he was four years old. He was a bright boy and eventually studied to be a doctor at McGill University and graduated in March 1876. While he was still a student, he had an affair with Flora Brooks and got her pregnant. He then aborted her baby, but was persuaded to marry her by her father who was a wealthy hotel owner.

The week after his wedding, Cream set sail for London to further his education and he enrolled at St Thomas's Hospital Medical School in London. There, he did postgraduate training but failed his exams, so he transferred to the Royal College of Physicians and Surgeons in Edinburgh, where he did qualify in 1878. He opened a practice in the town centre but one of his patients, a young woman, died under suspicious circumstances and Cream decided it was best that he return to Canada. Before this happened, he learnt that his wife Flora had died, apparently from tuberculosis, although it is thought that Cream sent her a letter containing a white powder which she was persuaded to take as a medicine.

Back in Canada, Cream soon found a new love, Kate Gardener. She too was soon pregnant but then was found dead in a shed in the alley behind Cream's surgery having apparently been overdosed with chloroform. Cream persuaded the coroner that it had been an accident and no further action was taken. Rather bizarrely, he then accused a local businessman of having made her pregnant and tried to blackmail him, but to no effect.

Cream now moved to Chicago, where he set up his practice in the red light district and found he could make money by performing illegal abortions. Again one of these went wrong and a client called Mary Faulkner suddenly died. This time Cream was arrested but escaped prosecution through lack of evidence. He told the coroner that his assistant was responsible for her death, and the case was not proceeded with. The following year another of his patients, Alice Montgomery, died – this time of strychnine poisoning – and, while Cream came under suspicion again, he was not prosecuted.

Apart from performing illegal abortions, Cream had devised another way to make money by selling a patent medicine for treating epilepsy. It contained a small amount of strychnine. When he advertised it, a railway worker of Boone County, Illinois, 61-year-old Daniel Stott, replied. He sent his attractive 31-year-old wife Julia to Chicago to pick up the medicine. Cream quickly charmed and seduced her, and they became lovers. She returned to her husband with a bottle of the medicine, to which Cream had added more strychnine, and her husband died soon after taking a dose. His death was attributed to his epilepsy, and he was buried without a post-mortem being carried out.

Cream had got away with murder yet again, but he then did something rather unexpected: he wrote to the local coroner accusing the pharmacist who had made up the medicine of adding too much strychnine to it. He called for Stott's body to be exhumed, which it was, and a Chicago analyst found 3.4 grains (220 mg) of strychnine in the stomach and 2.6 grains (150 mg) in the remains of the medicine bottle. Cream was arrested, put on trial, found guilty and sentenced to life imprisonment. He served his term in Joliet gaol until 1891, when the Governor of Illinois commuted his sentence to 10 years and he was released. It was later said that money had changed hands to bring this about, and by then Cream had inherited $16 000 on the death of his father (equivalent to $400 000 today).

Cream decided to return to England and arrived at Liverpool on 1 October 1891. He took a train to London and found lodgings at 130 Lambeth Palace Road, in the Waterloo area. He now called himself Dr Thomas Neill. One of his first purchases was strychnine and with this he poisoned two prostitutes, 19-year-old Ellen Donworth on 13 October 1891 and 27-year-old Matilda Clover on 21 October. It appears he persuaded them to take one of his capsules, which contained strychnine. Their deaths were suspicious but no one was arrested. After the death of Donworth, Cream wrote a letter to one of the partners of the book chain W. H. Smith & Son saying the partner was suspected of murdering the prostitute, and offering to act on his behalf. If he agreed to this, he should put a letter of agreement in one of the firm's bookshop windows, which he did, but Cream nevertheless did not contact him.

Cream next attempted to murder Louisa Harvey, who plied her trade at the Alhambra and St James's Music Halls. Cream took her to a hotel in Berwick Street where they spent the night together. He also told her that he had a cure for some spots she had on her face and agreed to meet her at Charing Cross Underground station the following evening. After having a glass of wine together, they walked on the Embankment where he gave her two capsules which she pretended to take but which she threw into the river. They had planned to go to the Oxford Music Hall but Cream now said he was unable to go and gave her five shillings (£75 in today's money) instead, and she went by herself. Cream had assumed she would soon be dead and he later confessed to her murder. When she read about this in the newspapers, she contacted the police and told them her story.

Cream decided to boost his income by selling patent medicines. He contacted the G. F. Harvey Company and negotiated to act as their agent in London. He returned to America where he purchased a large quantity of the medicine. He then moved on to Canada and booked a room at a hotel where he got into conversation with another resident, a Mr J. McCulloch. He showed him a bottle of a powder which he said could be used to procure an abortion, and that he gave it in the form of a gelatine capsule.

Cream returned to London in April 1892 and began to associate with prostitutes again. He met 21-year-old Alice Marsh and 18-year-old Emma Shrivell, who plied their trade at 118 Stamford Street near Waterloo Station. Cream later said that they charged him two shillings for their services (£30 at today's prices). Then, on 12 April he poisoned them, perhaps persuading them that his special capsules could be of benefit. However, he was seen leaving the house in the early hours by a passing policeman, PC Comley, who later was able to identify him. Soon the two women began to experience the symptoms of strychnine poisoning and called for help. A cab took them to hospital but Alice died before they got there and Emma died the following morning.

Cream now wrote a letter to the father of Walter Harper, who was lodging at the same address as him, and who was a student at St Thomas's Hospital. Cream tried to blackmail the man's father, saying that he knew his son had poisoned the prostitutes but would not inform the police if the father paid him £1500 (equivalent to around £500 000 today). Rather oddly, with the letter he enclosed a newspaper cutting about the earlier death of prostitute Ellen Donworth. The father informed the police, and Cream was arrested for attempted blackmail on 3 June 1892.

When the police searched Cream's rooms at the lodging house, they discovered a remarkable collection of bottles of pills, some of which contained strychnine. The body of Matilda Clover was exhumed from a common grave at Tooting Cemetery. A Dr Stevenson analysed the stomach, liver, and brain and found strychnine, and was able to extract enough of this toxin to kill a frog.

Cream's trial for murder began on 17 October 1892 and it was widely reported in newspapers. The jury took only 10 minutes to find him guilty but then there was a stay of execution to allow evidence of his insanity to be obtained from the USA. This was not sufficiently compelling to commute his sentence to life imprisonment in Broadmoor, the prison for the criminally insane, and he was executed on 15 November 1892. While in prison, he claimed to have murdered other women and even to be Jack the Ripper, although this was impossible as he was in prison in America when those crimes were committed.

9.7 MURDER, MICHIGAN, 1903

Mary Murphy was born in 1857, one of nine children of Irish immigrants to Canada who eventually moved to Michigan, where Mary trained as a nurse. She married a painter and decorator, James Ambrose, in 1876. James's business partner was Ernest McKnight and the two families lived together in the same house in the lakeside town of Alpena. Mary had five children, three of whom died as babies although two, Minnie and May, lived until they were two and five years old, respectively. Then, early in 1887, Mary's husband James died an agonising death and Mary collected $2000 insurance money (equivalent to around $140 000 or £100 000 today). There is every reason to believe that Mary poisoned him with strychnine.

Not long after Mary had been widowed, Ernest's wife became ill with episodes of convulsions and paralysis, similar symptoms to those exhibited by James on his death bed, and then she also suddenly died. Her sister came to look after her and she brought her baby boy along with her. He died.

It was at this time that Mary went with her own daughter May to visit friends at Saginaw, 150 miles from Alpena, but they both became ill on the train and were taken to hospital on arrival, where Mary recovered but May died, ostensibly of diphtheria. When Mary had recovered from her illness and coping with the death of her child, she returned to the McKnight residence and married Ernest. They moved to the village of Grayling, Michigan, where they lived for 10 years. Ernest died in 1898. He was taken ill soon after lunch one day, but recovered, only to die the following day. Mary collected another $2000.

Mary then went to live with her mother, Sarah Murphy, in the hamlet of Fife Lake, Michigan. A niece of Mary's, also called Mary, lived with them. In 1903, they were joined by Mary's brother John and his wife Gertrude and their baby Ruth. They had bought a plot of land nearby and were building a house. On 20 April, Mary was left babysitting while Gertrude went to help her husband. They returned to find baby Ruth dead. Mary said she had had a fit and suffocated in the bed covers. Gertrude was distraught, and John went off to inform the authorities and order a coffin for the baby. By the time he returned, Gertrude herself was also dead. Mary said she had a "seizure" brought on by her intense grief. A doctor was called, and he certified that her death had been caused by the shock of losing her daughter, although he did notice the curious arched posture of her body. Mother and daughter were buried together.

A week later, on 2 May, Mary's brother John also suddenly died. He too had been poisoned with strychnine, and we know that because Mary later confessed to his murder. Baby Ruth had been given it with a spoon, no doubt disguised with something sweet, while Gertrude was persuaded to take a capsule which Mary said would calm her nerves, and John was also given one of them. He died exhibiting the classical symptoms of strychnine poisoning and rumours began to spread that all the family had been poisoned. A few days prior to all this, Mary had bought five cents-worth of strychnine, ostensibly to kill some mice which she said were infesting the basement of the house.

What led to the deaths being investigated was Mary's subsequent behaviour. On 7 May, she went to the County Record

Office and presented an indenture for a debt of $200 that John had borrowed and which she now wanted repaid from his estate. This document had been drawn up four years earlier by a Justice of the Peace but it hadn't been officially recorded. Rather oddly, the debt had recently increased to $600 (equivalent to around $42 000 today, or £30 000). It all seemed rather suspicious and investigations began.

The local prosecutor ordered John's body to be exhumed, despite Mary's claim that she had promised him that she would never let this happen. His stomach was sent to the University of Michigan for analysis and strychnine was discovered. The bodies of Gertrude and the baby were also exhumed, and they too contained the poison. Mary was arrested on 31 May, and her trial began on 1 December 1903 at the Wexford County Court House, where she was charged with murder. Her lawyers offered a robust defence, claiming that John's death was either accidental or suicide, and the strychnine found in the other bodies was from embalming fluid. The jury members were not fooled, and Mary was found guilty and sentenced to life imprisonment, of which she spent 18 years in Detroit House of Correction. She was released on parole when she was 58.

Mary may well have murdered 15 people in total. In May 1892, a Mrs Eliza Chalker and her daughter had taken tea with Mary, but soon the child was twitching and foaming at the mouth and died four hours later. The following year, Mary's sister, Sarah Murphy, also died soon after taking tea with Mary. In 1896, another relative of Mary's called Mrs Carey died in rather mysterious circumstances. And, on Good Friday 1902, a child for whom Mary was babysitting started twitching violently and foamed at the mouth and died. In any event, it appears that Mary may hold the record for the number of strychnine murders.

9.8 MURDER, BYFLEET, SURREY, 1924

In January 1924, a 45-year-old French wireless operator, Jean-Pierre Vaquier, began a torrid affair with an English visitor, Mrs Mabel Jones, who was staying at the Hotel Victoria in Biarritz where he worked. She was convalescing from a nervous breakdown. Eventually, she had to go back to her husband, who was the landlord of the Blue Anchor hotel in Byfleet, Surrey.

Soon after Mabel returned to England, Vaquier followed, first staying at the Russell Hotel in London, and there she visited him and they renewed their intimate relationship – or so a chambermaid later testified. Vaquier then moved to the Blue Anchor itself, telling Mr Jones that he was in the UK on business. He deferred paying his bill for several weeks, saying that he was waiting for a cheque to arrive once he had completed a business deal. What he planned to do was to marry Mabel himself and this would involve disposing of Alfred.

Alfred was a heavy drinker and every morning he took a glass of health salts[††] to counteract his usual hangover. He dissolved a spoonful of this in water and gulped it down. It was to be the perfect vehicle for poisoning him. Jean-Pierre put some strychnine in the bottle of powder and the next morning Alfred duly consumed it, although he said to Mabel that the powder did not fizz in the usual manner and complained that it tasted very bitter. She even put a little of it on her tongue and confirmed it was so.

She suspected that Vaquier had put poison in the bottle but now realised what a dangerous position she would also be in if Alfred were to die. So she gave Alfred a mixture of salt and warm water to make him vomit and then some bicarbonate of soda and a cup of tea. But it was too late. Alfred began to shiver and went to bed. A doctor was called and by the time he arrived at 11:30 am, Alfred was experiencing violent convulsions and died soon afterward. His death was very suspicious and the police were sent for. Alfred's body was removed and a post-mortem was performed. Various organs were sent for forensic investigation in London. This was carried out by chemist John Webster of the Home Office, and he found 17/30th of a grain (37 mg) of strychnine.

The bottle of stomach mixture from which Alfred had taken the fatal dose was found to have been rinsed out, but there was enough strychnine in the dregs to identify the poison had been present. Vaquier was questioned but denied all knowledge of it.

Alfred had clearly been murdered with strychnine and Vaquier was the chief and only suspect, but where had he obtained the poison? His photograph appeared in a newspaper and the owner

[††]The best know version of this type of drink was Andrew's Liver Salts.

of the shop where Vaquier had bought the poison recognised him. He later picked him out at an identity parade, whereupon Vaquier greeted him! He said that Vaquier had claimed to need the strychnine for some experiments and was sold two grains (130 mg) of strychnine hydrochloride and signed the poison register under a false name.[‡‡]

This quantity of strychnine was insufficient to account for that detected in Jones's remains if it had been mixed with the health salts. A spoonful could not have delivered a fatal dose. The mystery of the extra strychnine came from Vaquier's attempt to blame someone else for Jones's death. He said that he had seen an unidentified woman hide a bottle behind a loose brick in the wall of an outbuilding at the Blue Anchor, and there the police found it. This contained 23 grains of the poison (1500 mg) but where Vaquier had bought it was never discovered.

Vaquier was tried for murder at the Surrey Assizes on 5 July 1924, found guilty, and hanged. A few days before he was due to be executed, he wrote to the police accusing Jones of running the Blue Anchor as a brothel and blaming others for his death.

9.9 MURDER, HAMPSHIRE, 1931

Frances Rollason had divorced her husband, George Jackson of the Royal Army Veterinary Corps, and then married 29-year-old Lieutenant Herbert ('Hugh') Chevis in December 1930. She was a 28-year-old wealthy heiress who had inherited around £200 000 when she was a baby (around £15 million today), and this was held in trust until she married or reached her 25th birthday. Hugh also came from a wealthy background, although not so wealthy as his wife.

Hugh had been born in India, the son of Sir William and Lady Chevis. He had been sent back to England to be educated at Charterhouse School and then went to the Royal Military Academy, from where he graduated in 1923 as a Second Lieutenant. In 1931, he was an instructor at the army's Aldershot Barracks in Hampshire, and he lived in a bungalow at nearby Blackdown Camp.

[‡‡]He signed the poison register as J. Wanker (*sic*).

On the evening of Midsummer's Day, 21 June, Hugh and Frances sat down to a meal of partridge, of which Hugh was particularly fond and which Frances had ordered at a London shop and had it delivered to their bungalow. The meal was prepared by their cook and served to them by Hugh's batman. When he ate his first mouthful of the meat, Hugh said it tasted so nasty that he rejected the rest of it. Frances said she had also tasted a little of the bird and decided it was bad. Hugh ordered his batman to take it back to the kitchen and dispose of it, and cook threw it on to the kitchen fire.

Soon after his meal, Hugh began to feel ill and was soon was suffering convulsions and was clearly in agony. A doctor was sent for and he diagnosed a severely upset stomach. An hour later, Frances too claimed to be feeling ill and a second doctor was sent for before the couple were rushed to Frimley Cottage Hospital. A few hours later, at 1 am on Sunday morning, Hugh died, despite the efforts of doctors who had given him a strong emetic to clear his stomach followed by artificial respiration. Frances quickly recovered and the next morning she returned to her London flat.

The partridge had come from Manchuria and one theory was that it had been poisoned there because strychnine-containing berries were sometimes used to kill these birds, which were then left to poison foxes and other vermin. Maybe one of these poisoned birds had somehow got into a consignment of game sent to the UK, but it seemed highly unlikely. Another theory was that the partridge had been poisoned while it was in a "meat-safe" container outside the back door of the bungalow where it had been hanging. Nor was it likely that the cook or the batman would have poisoned one of the birds. The only other suspect was Frances, which seemed unthinkable at the time. Perhaps she believed that strychnine could act as a sexual stimulant, a reputation which it had but was undeserved. Analysis of the gravy from the meal showed there to be two grains (130 mg) of the poison, and this must have been added after it had been taken into the dining room because the cook said she had tasted the gravy after she prepared it.

The murder of Hugh Chevis was to become even more mysterious following the arrival of a telegram sent from Dublin to Sir William Chevis on the day of his son's funeral, on

Thursday 24 June 1931. It simply read 'HOORAY, HOORAY, HOORAY!' and appeared to have been sent by a man called J. Hartigan, who was residing at the Hiberian Hotel in that city. However, enquiries by the police discovered that no one of that name had been a resident there. When enquires were made at pharmacies in the city, one of them had a record of selling strychnine to a man in June. His description matched that of the man who had sent the telegram.

But was the Dublin telegram just a press hoax, perhaps to keep a good story going? Or was it totally unconnected and sent by someone with a personal grudge against Hugh? The sender would have known of Hugh's death because it had been announced in *The Times* on the Monday. However, they would also have had to know Hugh's father's address of 14 Argyle Road, Boscombe, Hampshire, for the telegram to be delivered on the day of the funeral.

Despite a lot of effort by the police, no one was brought to justice for Hugh's murder. It seems likely that his wife was responsible, and her apparent symptoms of strychnine poisoning came on an hour after she had sampled the partridge, which suggests she was not affected and was feigning the symptoms to protect herself. We will never know. Frances eventually became something of a society hostess, and she subsequently married another three men and was twice divorced.

9.10 MURDER, LINCOLNSHIRE, 1934

Arthur Major was a 44-year-old lorry driver who lived at Kirkby-on-Bain in Lincolnshire,[§§] and who worked for a local quarry. He was married to Ethel Brown, the daughter of the gamekeeper on the nearby estate of Sir Henry Hawley. What Arthur didn't know was that when Ethel was 22 she had had an illegitimate baby girl, Auriel, who her parents passed off as one of their own.

Arthur was a soldier in World War I. He had been badly wounded and invalided out of the army. He met Ethel and they were married on 1 June 1918, and they lived with Ethel's parents. The following year, their son Lawrence was born. In 1929, the

[§§]Kirkby-on-Bain is a village 22 miles east of the county town of Lincoln.

couple qualified for a newly-built council house and it was soon thereafter that Arthur learned the truth about Auriel being his wife's child. He demanded to know who the real father was but Ethel would not say. (Indeed she never revealed who it was.) The marriage was effectively over. Arthur became violent towards his wife and began to seek consolation elsewhere. Ethel now schemed to get Arthur out of her life.

Ethel's first move was to write to the Chief Constable of Lincolnshire saying that her husband was using his lorry illegally to transport passengers and was often drunk behind the wheel. He should be locked up, she said. She also informed the village policeman of this, and he tried to discover whether this was so, but he found it was untrue.

Next, she thought of a way she might get a divorce and accused Arthur of having an affair with a neighbour, Rose Kettleborough. She produced a letter addressed to him from Rose, which she showed to a solicitor who then wrote a letter to Rose demanding that she stop the relationship. In fact, it transpired that Ethel had written the letter herself.

What Ethel didn't realise was that Arthur really was having an affair with another woman in the village, until a poison-pen letter informed her of this. Ethel now tried to get their council house transferred to her name and even forged a letter purporting to come from Arthur requesting this change. When the council wrote to Arthur, asking him to confirm this was what he wanted, he denied all knowledge of it.

Ethel decided that there was no way of getting Arthur out of her life but to murder him, and she would do it in a way that suggested he had died of natural causes. She started putting strychnine in his food in small doses. On the first occasion, she put a little in a corned beef sandwich she had made for his lunch. As he sat with a fellow worker one lunch time, he suddenly spat it out because it was so bitter and threw the rest of the sandwich to the birds. Those that ate it quickly died.

On 22 May 1934, Arthur came home from work and made his own evening meal from corned beef. He then went into the back yard with his son Lawrence to repair a bicycle but suddenly started trembling and almost passed out. Lawrence called his mother and together they got Arthur upstairs to bed. When Ethel's father called to see them, he discovered Arthur foaming at the mouth and twitching constantly. He immediately sent for the local doctor, despite Ethel's protests that this was unnecessary.

The doctor found Arthur suffering from convulsions and sweating profusely. Mrs Major told the doctor that she feared that the corned beef in the sandwiches she had made for his dinner had caused food poisoning, but said that he had been prone to fits for quite some time. The doctor diagnosed epilepsy and prescribed a sedative.

The following day, Wednesday 23 May, Ethel went round to the doctor to say that, although Arthur had had another fainting fit, he appeared to have recovered so the doctor didn't need to visit him again. The following evening, Arthur died at 10:40 pm, and on the Friday morning Ethel visited the doctor to say that her husband had had another attack and died in the night. The doctor issued a death certificate, giving the cause of death as an epileptic fit. Ethel now wanted his funeral to be held on Sunday 27 May, and when other family members questioned her haste, she told them that the vicar was going away the next day so it had to be that Sunday. But suspicions had been aroused and the funeral was not to take place.

Police in the nearby market town of Horncastle, five miles from Kirkby-on-Bain, had received an anonymous letter from someone who signed themselves as "Fairplay." It said that a few days before he died, Arthur Major had complained that his food tasted off and had thrown it out for a neighbour's dog to eat, and that this had promptly died. The writer of the letter was never identified.

Police started enquiries and soon learned from people living in Kirby-on-Bain that the Majors were a dysfunctional couple. Arthur spent most of his evenings in the village pub, and Ethel was known for spending recklessly, to such an extent that Arthur had placed a notice in the local newspaper saying he would no longer be responsible for his wife's debts.

The police ordered an autopsy, and they also dug up the poisoned dog and sent organs for analysis to Dr Roche Lynch of St Mary's Hospital, Paddington, London. Soon word came back that Arthur and the dog had ingested lethal doses of strychnine: 1.27 grains (80 mg) were found in Arthur's organs and 0.16 grains (10 mg) in those of the dog. Dr Lynch was also of the opinion that Arthur had imbibed strychnine on two occasions and had been given between 2–3 grains (130–200 mg) in total.

Mrs Major was questioned and said that her husband's last meal was of corned beef from a tin that their son had been sent to buy. She said it would be impossible to poison such a product, and in any case, there was no evidence that she had ever bought strychnine, and there was none in the house. Rather tellingly, she said this to the detective before details of which poison had killed him had been released.

The police soon discovered where the strychnine had come from: there was a bottle of it in a locked box at her father's house. He was justified in having strychnine in order to keep down vermin on the estate where he worked. He said there was only one key to the box and although he had once had two keys, one of these had been lost many years ago. A search of Ethel's home brought it to light.

Ethel Major was tried for murder at Lincoln Assizes in November 1935, and although she was defended by one of the country's most eminent barristers, Sir Norman Birkett, he could not save her. He claimed that she was regularly beaten by her unfaithful husband. Birkett decided not to put Ethel in the witness box, a decision which the judge adversely commented on several times in his summing up. The jury took only an hour to find her guilty but recommended mercy because they believed she really was a battered wife. Despite their plea, the Home Secretary, John Gilmour, did not find any extenuating circumstances to justify a stay of execution, and she was hanged at Hull Prison at 9 am on 19 December 1935.

9.11 ATTEMPTED MURDER, SPITZ-AN-DER-DONAU, AUSTRIA, 2008

On 8 April 2008, Hannes Hirtzberger, the mayor of the north Austrian town of Spitz-an-der-Donau, returned to his parked car to find some Mon Cheri liqueur praline chocolates attached to his windscreen wipers with a note that read "I want to tell you something. You are someone very special to me." The chocolates contained strychnine. The following day, the mayor ate one of them in his car as he drove to work and then suddenly became ill. He pulled to the side of the road and rang for help. The paramedics who came to his aid managed to save his life but he was to be left permanently affected because the strychnine had also triggered a heart attack.

In fact, it transpired that the chocolates had been the "gift" of 56-year-old Helmut Osberger, a local inn keeper and wine producer. His reason for attempting to murder the mayor was that he assumed he was the man responsible for blocking Helmut's planning application to develop an old vineyard on which he wanted to build an hotel, or at least sell it to developers who would do this.

The mayor had asked Helmut for more information about what he intended to do with the land when he had permission to develop it, but Helmut was reluctant to divulge this and blamed the mayor for blocking his application.

Police investigations eventually pointed to Helmut as the likely poisoner, and indeed, his DNA was found on the inside of the chocolate wrapper. When the police asked Helmut for a saliva sample of his for testing, he tried to submit a sample of one of his son's saliva instead.

Helmut was put on trial in Krems, a nearby town, in December 2008 and was found guilty and given a sentence of 20 years in prison.

9.12 MYSTERIOUS SUICIDE, SADDLEWORTH MOOR, LANCASHIRE, 2015

At 10:47 on the morning of Saturday 12 December 2015, Stuart Crowther was cycling in the Peak District National Park when he passed a man who was lying in the grass at the side of the road. Naturally, Stuart stopped to see what was the matter, only to discover that the man was dead. It appeared he had been out walking and possibly suffered a heart attack. He clearly was not a hiker because he was dressed in casual clothes. Crowther called for help.

The body was that of a tall, white, elderly man – he was 6 ft 1 in – possible aged around 70. He had blue eyes, a receding hairline, and a large nose. But who was he? There was nothing in his pockets to identify him – no wallet, credit cards, or mobile phone – only £130 in cash and some train tickets, but these were enough for his movements the previous day to be retraced, and his image was caught on CCTV at several locations.

The first of these showed that he had started his journey at 9 am at Ealing Broadway underground station in London. He was next seen at London Euston station, where he bought a return ticket to Manchester for £81, and he caught the 10:00 train, arriving there at 12:07, where he was again captured on CCTV.

He remained in the precinct of the station for almost an hour, visiting various shops and buying lunch before moving off towards the city centre. Thereafter, he was off-camera so we do not know what he then did, but he turned up mid-afternoon at the Clarence Pub in Greenfield, Saddleworth. There, he asked for directions to Indian's Head. The landlord of the pub, Melvin Robinson, later said he had a Northern accent and advised the man not to visit the moor that day as it was raining heavily and would soon be getting dark. However, the man set off and was seen walking up the hill at 4:30 pm by two birdwatchers. A few hours later, he died what must have been an agonising death from strychnine poisoning. What led him to do such a thing? Perhaps he did not realise what a cruel poison he had taken. He did not leave a note explaining what had driven him to take his life this way, nor why he wished to die in this particular location.

Indian's Head is an outcrop on Saddleworth Moor near Dovestone Reservoir. This had been the scene of an air crash on 19 August 1949 in which 24 passengers and all three members of the crew on board a Douglas Dakota BEA flight were killed. Eight people survived and one of them is still alive: Professor Stephen Evans. He told reporters how his younger brother died but he and his parents survived. Also, a two-year-old baby, Michael Prestwich, survived but he was killed 10 years later in a train accident. Was the unknown man visiting the scene of the accident for some personal reason? However, he would have been only a baby when it had occurred 66 years ago. Yet clearly he wanted to die in this particular location and had travelled a long way in order to do so. He was well dressed with a white shirt, blue cords, blue jumper, brown jacket – all from Marks & Spencer – and black Bally shoes which cost around £250.

The man brought the strychnine with him, and it was dissolved in water in a plastic bottle that previously had held sodium thyroxine, a drug that is prescribed for those suffering from an underactive thyroid. It had been produced by the GlaxoSmithKline pharmaceutical company and its label was also printed in Urdu, which showed that its original contents had been manufactured in Pakistan. There it is still possible to purchase strychnine.

Police enquiries failed to identify the man and no one in Ealing reported him missing, and it was not until January 2017 that someone came forward to reveal his identity. A former neighbour in the London Borough of Streatham showed the police a video they had of a local wedding of 1994 in which a man they said was David Lytton could be seen dancing with his partner. Indeed the woman had been his partner for more than 15 years. A DNA match with a relative of Lytton's confirmed his identity.

Lytton was the grandson of a family of Jewish refugees who came to the UK from Poland in 1901. They were called Lautenberg but changed their name to Lawson, and David later changed this to Lytton.

For many years he had worked as a driver on London Underground and then got a job as a croupier in a casino, until he had been made redundant. He then began taking in lodgers, one of whom was from Pakistan and with whom he struck up a close friendship.

In 2006, and no longer able to keep up the mortgage payments on his house, he left his partner and moved to Florida for a while. Then he went to live in Pakistan and remained there until he returned to the UK to die.

The reason for committing suicide in the place he chose, and why he did so, remains a mystery that will probably never be solved. Members of his family and a former girlfriend said Lytton was often depressed and rather a loner who was difficult to get to know.

CHAPTER 10

Digitalis and a Mysterious Death in Verona in 1329

A word in **bold** *indicates that further information can be found in the Glossary. Only the first time the word appears in a chapter will it be so indicated.*

Digitalis purpurea is the botanical name of a plant we generally call a foxglove. Foxgloves prefer acidic soils and shade and are often found in woodland. In the first year of growth, the plant produces only leaves and it is during the second year that it produces the flowers from which it gets its name. The one with pink flowers is the most common kind found in the UK (Figure 10.1). There are other varieties in other parts of the world, and all share a rather unexpected characteristic that conflicts with their attractive appearance: their leaves are highly toxic because the plant produces the deadly poison digitalis. Deadly though it is, for more than 200 years this chemical has been used to improve lives blighted by heart disease.

Perhaps because it still features in some crime novels and TV dramas, digitalis is seen as an easily available poison if you have access to foxglove leaves. A recent case was that of Lisa Allen of Denver, Colorado. She agreed in April 2010 to plead guilty to attempting to poison her husband with a salad which contained foxglove leaves, and she was sentenced to four and a half years in

More Molecules of Murder
By John Emsley

Published by the Royal Society of Chemistry, www.rsc.org

Figure 10.1 Foxglove plants contain digitalis, which acts as an acute diaretic. © Michael Warwick/Shutterstock.

jail. Her husband had noticed that the salad tasted rather bitter but assumed his wife had used some new herb that had become fashionable. He subsequently needed hospital treatment for digitalis poisoning, but survived.

10.1 DIGITALIS

Digitalis used to be prescribed as a healing drug, but it was always known that it could be fatal if misused. It consists of two almost identical alkaloid chemicals: **digitoxin** and **digoxin**. Digitoxin is the main one in *Digitalis purpurea* while the almost identical molecule digoxin (pronounced dy-jox-in) predominates in *Digitalis orientalis.* The first person to isolate digitoxin was Oswald Schmiedeberg in 1875, but it was not until 1925 that a chemical structure was suggested by Adolf Windaus and this was not confirmed until 1962.

Both of these chemicals have two modes of action on the heart: they make the heart's muscle action stronger and they reduce the strength of the electrical signals that govern the motion of the heart, especially if these have become disorganised causing irregular pulses (fluttering). Too high an intake of digoxin will supress the signals altogether and that will result in death within minutes.

Digitoxin and digoxin are readily absorbed through the stomach wall, or they can be injected and then they work immediately. These molecules work by interfering with the sodium mechanism by which this essential element moves into cells, thereby increasing the amount of sodium in the fluid between cells. This in its turn increases the calcium being pumped into cells and this has an effect on the way the heart muscle moves, to the extent that it is able to pump more blood. These molecules do this *via* their effect on the enzyme Na^+/K^+ATPase.[†] However, this enzyme has other roles to play in the body and blocking it causes a feeling of nausea and loss of appetite, which is why digitalis was once thought of as a slimming aid. The enzyme is also a key part of the retina of the eye, which is why digitalis can make it seem as if you are viewing everything through a sheet of pale yellow glass or see haloes surrounding points of light. Vincent van Gogh's so-called "yellow period" may be linked to his treatment for epilepsy in 1888 when he was prescribed digitalis. It was then that he produced his most famous pictures depicting sunflowers. His paintings of the sky at night exhibit the halo effect. When he painted two portraits of his doctor, Paul Gachet, he included foxgloves in both.[‡] However, digitalis may also have made him very depressed.

The treatment of acute digitalis poisoning should begin with an emptying and washing out of the stomach, which might have already begun with vomiting. The next step is to give the patient a lot of potassium chloride (KCl) dissolved in fruit juice, but this must be carefully done because too much potassium in the body can also be toxic. This advice has been revised in recent years because the use of potassium chloride in fruit drinks is seen as possibly causing a heart attack (see Chapter 6).

Current treatment involves the intravenous administration of antibody fragments of digoxin and the careful management of electrolytes. The amount of potassium in the blood may go either way, *i.e.* hyperkalaemia or hypokalaemia, but only the latter

---

[†]This enzyme, as its name implies, controls the movement of sodium ions (Na^+) and potassium ions (K^+) in and out of nerve cells and as such are the way in which electric pulses move along nerve fibres.

[‡]One of these pictures, painted in 1890, was bought by a Japanese paper manufacturer, Ryoei Saito, for $82.5 million in 1990, only to be sold back to the auctioneers Christie's for $10 million a few years later.

would be treated by careful infusion of potassium chloride. Slow heart beat, *i.e.* bradycardia, would be treated with **atropine**, whereas a fast heart beat, in which this exceeds 100 per minute, *i.e.* tachycardia, would be treated with phenytoin or lidocaine.

Finally, there are other drugs, such as cholestyramine, which can be given to speed up the metabolic disposal of digitalis and thereby shorten the half-life of the drug within the body.

The symptoms of digitalis poisoning were once used to cheat insurance companies, and in New York in the 1930s, a group of doctors and lawyers did this by using it to mimic heart disease. They could show from cardiograms that someone was unable to work and so could claim money to support them. Eventually, the insurance companies became suspicious, and in 1935 they sought legal advice and eventually obtained evidence of what was going on by tapping telephone conversations. Some of those involved confessed after hearing the recording of what they had said, and eventually 75 people were brought to trial and found guilty of fraud. Their lawyers challenged the telephone evidence as illegal and this went to the Supreme Court, which upheld it as acceptable. A Professor of Medicine who had validated the cardiograms committed suicide.

Cases of murder with digitalis also occurred in fiction, and it was the poison used in Agatha Christie's *Appointment with Death*, published in 1938. The story is set in Palestine, and concerns one member of a family party visiting the ancient city of Petra being murdered having been injected with digitalis. Hercule Poirot, who was holidaying nearby, quickly solved the case.

10.2 THE DOCTOR WHO DISCOVERED THE HEALTH BENEFITS OF DIGITALIS

William Withering was born in Wellington, Shropshire, in 1741. He came from a family of doctors and surgeons, and his father ran a pharmacy. William decided he too would like to be a doctor, and he obtained his degree at the University of Edinburgh Medical School in 1766. He then spent time in France before working first at Stafford before moving to Birmingham. It was there that he learnt of a herbal remedy from a patient who was suffering from dropsy. This is the condition in which parts of the body, such as the legs, swell up due to fluid retention.

Withering had been unsuccessful in treating the man, who then turned to traditional medicine – and it worked. He told Withering what he had taken and the doctor decided to investigate.[§] He found that the remedy included various herbs, but in particular, it contained foxglove and he judged that this was the active ingredient. He then heard that a friend of the family, Dr Ash of Brasenose College Oxford, had been cured in a similar way and so he contacted him. He encouraged Withering to begin a series of experiments with foxglove. This he did and in a way that has since been regarded as the first systematic testing of a drug.

Withering carried out a series of trials on his poorer patients who suffered from dropsy. He asked them if they would volunteer to take part, which they were pleased to do, and he eventually wrote an account of the doses of digitalis he had used and the effects this had had. He even included in his account those patients who had not benefited. However, most of them had got better, and Withering is still remembered as a pioneer in drug research.

He published his findings and test results in his 1785 book entitled *An Account of the Foxglove and Some of its Medical Uses*. Withering described the effects that foxglove extract had on 134 patients given digitalis. For example, Patient No. 20 had swollen legs and was producing only tiny amounts of urine. When he was given digitalis, it made him sick but acted as a powerful diuretic, and within a few days, it had cured his swelling and his output of urine returned to normal. The same happened with Patient No. 81 who was 33 years old and who liked his ale, but whose belly and legs had become bloated. Ten days of digitalis treatment restored him to good health. However, sometimes digitalis did not work, as happened with Patient No. 14, who not only had swollen belly, legs, and thighs but also suffered from gout. He was seriously ill, and digitalis had no effect on the man who died a few days after Withering began to treat him.

[§]Later accounts of his discovery report that the herbal remedy had come originally from a so-called Mother Hutton who lived in Shropshire, but this name was invented in the 1920s by an advertising agency for a company which was producing digitalis for medical use.

Withering died in 1799 aged 54 and his researches into digitalis came to an end. He died of tuberculosis and fittingly his tombstone had a foxglove carved on it. Not everyone agreed with Withering's findings, but many doctors who prescribed it said it transformed the lives of their patients; however, other doctors said it failed to cure.

When the body retains fluid, it is now referred to as oedema, and this may be due to a variety of complaints such as the inability of the heart to pump blood effectively or the malfunctioning of the liver known as cirrhosis. The action of digitalis will correct the former of these but will do nothing to aid the latter, and in 1827, a Dr Richard Bright published his researches showing this to be so. It has even been suggested that the digitalis prescribed for the English statesman Charles James Fox hastened his demise in 1806. He suffered from dropsy but it was probably caused by liver disease, and so prescribing digitalis would not solve the problem and may even have hastened his end.

Digitalis was used by doctors for other ailments for which it could have no benefit, and its reputation suffered. Its popularity gradually waned in the 1800s but it was still available for doctors to prescribe. Medical interest in the drug was revived early last century when its mode of action became clearer. This was brought about by the introduction of the electrocardiograph in 1903, which gave information about the state of the heart.¶

Originally, digitalis was prescribed as the powdered dry leaf. It would first be given in a relatively large dose for rapid effect, to be followed by a much smaller daily maintenance dose. Even so, it carried a risk – as later analysis showed. In 1961, a study of patients receiving digitalis in hospital revealed that 20% developed toxic symptoms and 11% died. Digitalis was at times prescribed for conditions other than heart disease, such as epilepsy and seizures, but it is no longer an accepted treatment for these conditions.

Digitoxin is extracted from the dried leaves of *Digitalis purpurea* by means of a 50/50 solution of ethanol and water.

¶The electrocardiograph was invented in 1887 by Augustus Waller of St Mary's Medical School in London, but it was developed by Willem Einthoven in 1895 and made commercially viable in 1903. Einthoven got the 1924 Nobel Prize in Medicine for his work.

Ten kilograms of leaves will produce around six grams of digitoxin. The measure of its toxicity was done using the **LD_{50}** method and shown in guinea pigs to be 60 mg per kg, but for cats it is much more toxic with an LD_{50} of 0.2 mg per kg. If this were the same for humans, then a dose of 14 mg could prove fatal, whereas on the basis of the LD_{50} for guinea pigs, it would require 4.2 grams.[‖]

Digitoxin was the main component of the digitalis originally used, but it was to be superseded by digoxin. Digitoxin has a half-life in the body of six days whereas digoxin's half-life is between one and two days. Moreover, digitoxin is more toxic than digoxin. Digitoxin, which is the main component of digitalis, was essentially the form prescribed until the 1930s when a Dr Sydney Smith, who worked for the drug company Burroughs Wellcome, first isolated digoxin. He extracted that chemical from the Balkan foxglove, *Digitalis lanata.*

His investigation came about as a result of an observation made by a Dr Frank Wokes in 1929 that the digitalis he extracted from some foxgloves was four times more potent than the digitalis extracted from the common foxglove. *Digitalis lanata* is now used as a source of digoxin, and in 1998 this chemical was reaffirmed as an approved drug by the Food and Drug Administration of America. Nevertheless, its use is declining because there is some evidence that it increases the risk of death among women to whom it is prescribed, but not when prescribed to men.

Digoxin is marketed under a variety of trade names, most of which indicate a link to the drug by incorporating 'go' or 'ox' in the name. In the UK, it is known as Lanoxin but in other countries it is called Cardioxin, Digomal, Toloxin, *etc.*

Digoxin is on the World Health Organization list of approved medicines that are considered to be an essential part of a basic medical health system. It can be prescribed for someone who is suffering from heart flutters, although generally a beta-blocker is the drug of first choice because there is less risk of an adverse response. Digoxin enables more blood to be pumped with every

[‖] E. Weinhouse *et al.* 'Comparison of digoxin-induced cardiac toxicity in resistant and sensitive species,' *J. Pharm. Pharmacol.*, 1983, **35**, 580–583.

beat of the heart. Digoxin may also be used when an abortion is performed to ensure that the foetus is not born alive.

Deaths during treatment have been caused through poor quality control during manufacture of the drug. In 2008, an Icelandic-based drug maker of generic digoxin tablets produced some that contained twice as much of the active ingredient as required for a normal prescription. It produced this under the trade name of Digitek, and when they realised what had happened, they informed the FDA who issued an alert, but not before some people prescribed the drug had taken it and suffered toxic symptoms. The following year, another generic drug maker, Caraco, based in Detroit, announced a recall of its product because of what it described as "tablet-size variability."** In Belgium, in 1978, a company manufacturing tablets mislabelled digitoxin as estradiol benzoate, which is a treatment for prostate cancer. As many as 40 patients may have died as a consequence. Some of their corpses were exhumed and shown to have the drug in their system. Another error occurred in Holland, in 1970, when 200 000 tablets were erroneously labelled as digoxin instead of digitoxin. Later, it was shown that this error may have contributed to the deaths of 19 people.

When digitoxin or digoxin is detected in the blood of someone who has died, and it is at levels above two nanograms per millilitre, then this indicates that a fatal dose had been given to that person. As prescribed by a doctor, the level of these in the bloodstream would be half this or less, but even so, the difference between safe and dangerous is a narrow margin.

And there have been deliberate killings with digitalis, and even murders committed with it centuries ago have recently been investigated.

10.3 ASSASSINATION, VERONA, ITALY, 1329

Can Francesco della Scala was born in 1291, the third son of the then ruler of Verona, Alberto I, one of the Scaligeri dynasty who controlled the city from 1277 to 1387. However, the third son is better known as Cangrande†† because of his triumphs on the

**The company was eventually taken over by Sun Pharmaceuticals and the plant was closed down in 2014.

††Cangrande means big dog.

battlefield and the mercy he showed to those he conquered. He was also a generous patron of the arts and eventually had poets and painters resident at court. The great Dante was a guest from 1312 onwards and praised him in his famous work *The Divine Comedy,* the epic poem that he had started writing in 1308 and which took 12 years to complete.

Cangrande was his father's favourite son, and he was knighted when he was only 10 years old, and just before his father died. Alberto I was succeeded by his eldest son Bartolomeo, but he only reigned for three years, to be succeeded by the second brother Alboino who reigned in partnership with Cangrande. In 1311, when he was only 20, Cangrande became the sole ruler of Verona.

Italy in the 1300s was a collection of small states controlled by various families who were often at war with one another. So it was that Cangrande spent much of his reign making alliances with various groups while fighting others, and he was often successful as a warrior; for example, he succeeded in conquering Vicenza in 1314. Despite some setbacks, he continued his successes and finally captured Padua in 1328 after a long and bitter struggle, and Treviso in 1329. He was then designated Imperial Vicar of Mantua by the Holy Roman Emperor.

On 18 July 1329, Cangrande celebrated his success at dominating all of Northern Italy by having a triumphal entry into Treviso. Soon thereafter he became ill and died on the morning of 22 July, his death being attributed to drinking water from a polluted spring. He was given a splendid funeral and his mummified remains were laid to rest in a monumental marble tomb in the church of Santa Maria Antiqua. However, there were those at the time who suspected he'd been poisoned and there were clearly some who benefited from his death, not least his younger brother, Alberto.

On February 2004, Cangrande's tomb was opened so investigations of his corpse could be undertaken. The mummy was still wearing the magnificent garments in which he had been buried. Samples of the hair, liver, and faeces from the rectum were removed for analysis and these revealed the presence of digoxin and digitoxin. It appears that during his last hours he had consumed a meal which included enough foxglove to provide him with a fatal dose. This was consistent with the symptoms he displayed during the last day and night of his life, which included vomiting, diarrhoea, stomach pains, and fever, easily mistaken by doctors for gastroenteritis.

The question of whether his death was due to a deliberate act to feed him a meal which included foxglove components, or was just an accidental death due to the misuse of foxglove leaves cannot be known, but it was known that this plant was dangerously poisonous. Foxgloves are mentioned in a manuscript written in 1120 at Bury St Edmunds in England, and called *Herbarium Apuleii Platonici.* It reported the use of foxglove extracts for treating wounds but warned that it was also very poisonous.

Reports written after the death of Cangrande claimed he was murdered by poison, and one account said this had been delivered *via* a fruit dessert, in which case it must have contained powdered digitalis, suggesting the work of a professional assassin. This is also the conclusion that is reached in the 2015 paper in the *Journal of Archaeological Science* (volume 54, pages 162–167)[‡‡] by the group of scientists who carried out the analysis of his body. The principal suspects are named as his ambitious nephew Mastine in conjunction with his brother Alberto, who succeeded Cangrande as ruler of Verona. Or it might have been the work of an assassin employed by either Milan or Venice, whose rulers were worried about the expanding power of Verona.

10.4 MURDER, PARIS, 1864

Couty de la Pommerais was a 24-year-old doctor who came to Paris in 1859. One day, he was consulted by a young artist, Mme de Pauw, who had recently been widowed. They became lovers, but it did not last. Pommerais was in debt due to gambling and needed another source of income, so he married the daughter of a wealthy widow. His mother-in-law died soon after dining with her daughter and son-in-law, and they inherited her estate (Pommerais certified that she had died of cholera although almost certainly he poisoned her with digitalis).

[‡‡]I am most grateful to Dr Simon Cotton of Birmingham University Chemistry Department who brought this work to my notice.

Meanwhile, his former lover Mme de Pauw was now penniless and approached Pommerais for help, and they renewed their affair. The doctor devised a way for her to solve her financial difficulties. She would take out insurance for 500 000 francs[§§] using money that she borrowed from Pommerais, and then they would fake an accident that would make it appear she was permanently injured and so the policy would make regular payments for her maintenance for the rest of her life. Pommerais gave her the money to make the first few payments on the policy and she signed a letter saying she was in debt to him for 100 000 francs.

The plan they devised was for her to appear to take a terrible fall down a flight of stairs, and this would be heard, if not actually witnessed, by other tenants in the block where she lived. Pommerais would then certify that she was so badly injured that it was unlikely she would ever walk again. He then said that the insurers would exchange her life policy for a much less costly invalid policy that would be paid annually for the rest of her life.

However, that was not what he'd planned. Suddenly, Mme de Pauw was taken very ill with copious vomiting and she died. He wrote on her death certificate that it was due to cholera. Now Pommerais came forward with the letter of debt, which had been changed to 500 000 francs. Her sister, who had been informed of the insurance scam, was naturally suspicious because the insurance money would not end up with the family as she was expecting, and she wrote a letter to a Chief Inspector Claude telling him what had happened. He began an investigation. He ordered the exhumation of Mme de Pauw's body. He also carried out an inspection of the contents of Pommerais' surgery, from which various bottles and letters were removed.

The inspector then sent samples from the corpse and the bottles for forensic analysis to France's leading expert, Professor Ambroise Tardieu. He quickly determined that the cause of death had not been due to a metallic poison like arsenic or antimony. Then, he looked for a plant-based poison and in particular digitalis, which Pommerais had prescribed for Mme de Pauw and which she mentioned in one of her letters, found in

[§§]The annual salary at the time for an average worker was around 2000 francs.

the surgery. The police had also discovered that Pommerais had recently purchased three grains (200 mg) of digitalis.

Tardieu had devised a new method to show the effect of digitalis. He took a frog, opened its chest and observing its beating heart. He then injected digitalis into the animal and noted its effect. Its heart beat slower and slower and eventually stopped. When Tardieu took another frog and did the same with material extracted from Mme de Pauw, it behaved in just the same way. What he also needed were samples of her vomit but none of these had been retained. However, the floorboard of Pauw's bedroom had gaps, and when these were removed, they were able to retrieve samples of vomited material in which Tardieu detected the presence of digitalis by the use of a frog's heart. When the case came to court, Pommerais' defence lawyers argued that this forensic test did not prove that digitalis was the cause of Mme de Pauw's death. However, the other evidence of motive and opportunity was enough to find him guilty, and he was executed by guillotine in June 1864, aged 29.

As for Tardieu's forensic evidence, his frog method continued to be used for many years to indicate digitalis, until reliable methods of chemical analysis were developed.

10.5 MULTIPLE MURDERS, LIEGE, BELGIUM, 1936

Marie Becker was born in 1877 in Liege, Belgium. She married a cabinet maker and led the life of a housewife until she was 55 years old, when she became an aged nymphomaniac and poisoner. In 1932, she was attracted to a younger man, Lambert Beyer, with whom she began an affair after murdering her husband with digitalis. However, when she discovered Beyer was unable to satisfy her sexual demands, she disposed of him in the same way in November 1934. Thereafter, she liked to spend her evenings in night clubs and dance halls and enticed those she fancied back to her home to share her bed, often paying them for their services.

To finance her lifestyle, Marie opened a fashionable dress shop using her husband's life insurance to set this up. However, it did not generate enough income to satisfy her needs. She began to befriend wealthy customers, and she would dispose of them by offering them a cup of tea while they looked at the latest fashions. This invariably made them ill, and so she would close the shop and take them back to their home, where she would take care of them while they were ill – but, inevitably, after a few days they would die. Then, she would rob them of what she could lay her hands on, including expensive clothes, many of which were eventually found at Becker's home. At the time, these deaths were invariably put down to natural causes such as a heart attack. Marie often went to their funerals and appeared grief-stricken at the loss of a seemingly good friend.

One of Marie's long-standing friends rather envied her lifestyle, and she was told that a similar lifestyle could be hers if only her husband were dead. Marie told her to try digitalis and even offered her some from her own cache of the poison. Her friend was amazed at what she heard, but it worried her so much that she went to the police and told them what Marie had done. They already knew of Marie and her lifestyle because they had received anonymous letters accusing her of causing suspicious deaths. She was arrested in October 1936.

The body of Marie's husband and various other victims were exhumed and traces of poison were found. The police discovered a bottle of digitalis among her belongings. She was put on trial and found guilty of causing the deaths of 10 people – some thought this number was more like 20 – and she died in prison a few years later. (The death sentence had, by then, been abolished in Belgium.)

10.6 MULTIPLE MURDERS, NEW JERSEY, 1987–2003

Charles Cullen was born in 1960 in West Orange, New Jersey, USA, the last of eight children. He is credited with killing more than a hundred patients during the years in which he was employed as a nurse at various hospitals in his home state and in Pennsylvania. Throughout the 1980s and 1990s, he definitely murdered 40 of his patients by over-prescribing digoxin. He admitted this in December 2003, and in March 2006 he was sentenced to 28 consecutive life sentences without parole, meaning he will not be released for another 400 years.

Cullen's career began with his enlisting in the US Navy in 1978, when he became part of the submarine service and served on the USS Woodrow Wilson, a submarine equipped with

Poseidon nuclear missiles. But the navy was not for him. He showed signs of mental instability and was discharged in 1984 on medical grounds.

He then enrolled as a trainee nurse in Mountainside Medical School of Nursing in New Jersey, from which he graduated in 1987. He got a job in the St Barnabas Medical Center, where he began his career as a mass poisoner. His first victim was Judge John Yengo, but there were several other unexplained deaths in the wards at the time. An internal investigation pointed the finger at Cullen as the person most involved in this curious state of affairs, and he was asked to leave. This sequence of events was to be repeated in the years that followed.

Cullen then moved on to the Warren Hospital in Phillipsburg, New Jersey, where he murdered elderly women by overdosing them with digoxin. While there, he took time off work and was treated for depression and he even attempted suicide. Then he got a job at the Hunterdon Medical Center at Flemington, and he seems to have given up his career as a poisoner for a few years, but began again in 1996. In 1997, he again was unemployed but eventually found work in a psychiatric unit at the Liberty Nursing and Rehabilitation Center in Allentown, Pennsylvania, where he continued to murder patients, often with digoxin.

His next move was to the Somerset Medical Center in Somerville, New Jersey, where he killed at least eight patients. It was at that hospital where staff became suspicious because Cullen was entering areas to which he was not assigned and accessing drugs that he was not authorised to use. Eventually, he was arrested and charged with murder.

How was Cullen able to operate for so long? The main reason was that he easily found a job as a nurse because they were in short supply and checks were not made with his former employers. Also, the people he murdered were often seriously ill so their deaths were not unexpected. Cullen said he was vague about many of the deaths that had occurred when he was on duty because he said that he lived his life "in a fog."[¶¶]

[¶¶]During his time in prison Cullen donated a kidney to save the life of the brother of a former girlfriend.

CHAPTER 11

Curare and the Psychopathic Surgeon from Argentina

A word in **bold** *indicates that further information can be found in the Glossary. Only the first time the word appears in a chapter will it be so indicated.*

Curare is not a poison that will kill you if you eat or drink something which contains it, but it will kill you if it directly enters the bloodstream. It is a muscle relaxant and, as such, will stop the lungs working, causing death by asphyxiation. Deadly as it is, curare had an important role in surgery because it acted as a muscle relaxant, but only in the minutest of doses.

In Agatha Christie's *Murder Ahoy*, a woman appears to be dying almost as soon as she injures her hand on a mousetrap that has been coated with curare. Another murder mystery in which curare featured was in Arthur Conan Doyle's *The Adventure of the Sussex Vampire*, when Sherlock Holmes and his friend Dr Watson explain why a mother was biting her child's neck, not as a vampire, but as a means of sucking out the poison from a dart which the child's elder brother had fired.

Curare may be deadly, but it is not the kind of poison that a murderer would generally use because it has to be injected into the body to kill; although, once this has happened, then death will follow within about 10 minutes. Naturally, very few examples

More Molecules of Murder
By John Emsley

Published by the Royal Society of Chemistry, www.rsc.org

of murder with curare have been reported, but it featured in a series of unexplained deaths in the USA in the 1960s and 1970s, and the person who used it was a surgeon: Mario Enrique Jascalevich. He realised there was an easy way to inject it without arising suspicion.

11.1 POISONED DARTS AND ARROWS

The natives of South America long ago discovered how useful curare could be when it came to hunting. They obtained it mostly from the bark and roots of a vine, the South American *Chondrodendron tomentosum*, although some was also extracted from the climbing shrub *Strychnos toxifera.* The natives of Brazil, Ecuador, Colombia, and Peru smeared it on the tips of arrows and on the darts fired from blow pipes. The arrows and darts had grooves cut into them and that is where the curare was placed so that it would not be wiped off as it penetrated the animal. And even if the animal they were hunting was only slightly injured, it would soon be disabled once the curare got into the bloodstream, where it would eventually paralyse the animal's lungs. They also knew that animals which were hunted and killed this way could be safely eaten (Figure 11.1).

Sometimes humans were the target, and the Spanish explorer Francisco de Orellano reported that one of his party had died as

Figure 11.1 Curare is used by hunters in South America as a poison on the darts used to kill their prey. The prey can be safely eaten. © Ammit Jack/Shutterstock.

a result of an arrow hitting his finger, and then not particularly deeply, but it conveyed enough curare into his body to kill him.

Tales of the remarkable effect of poison darts were soon circulating in Europe, thanks mainly to a book, *De Orbe Novo Decades*, written by the Italian Pieta Martyr D'Anghera (1457–1526).[†] In it, he mentioned the use of poisoned arrows by South American Indians and says that they killed 47 of a party of Europeans by this means. D'Anghera himself may even have been part of an expedition to the New World as a missionary for the Catholic Church, and he was commissioned to write about his visit as a member of the so-called Council of the Indies. It was he who first reported the effects which curare could cause, and noted that although an arrow might produce only a slight wound and barely penetrate the skin, the victim would die within a short time. Other explorers such as Sir Walter Raleigh (1552–1618) told of arrow poisons, and in 1595 he brought a sample of curare to England.[‡]

11.2 THE INVESTIGATION OF CURARE

Scientific interest in curare had to wait another 200 years. The first people to discover how the poison was produced were the German explorers Alexander von Humboldt and Aimé Bonpland, who sent samples of the sticky paste back to Germany for scientific examination. In fact, different tribes in different regions produced the toxic material in different ways but most came from the bark of *Chondrodendon tomentosum*, and in some cases, it was mixed with other toxins such as snake venom. The natives would crush the stems and roots of the vine and boil them in water, eventually reducing the resulting solution to a thick paste.

Curare was used in experiments at the University of Leiden in 1740, although the results of these were not published. At about the same time, an Englishman, Dr Brocklesby, also undertook some tests with it and noted that when he injected it into the leg of a cat, the animal stopped breathing, although its heart continued to beat for more than an hour afterwards. This proved

[†]In 1555, a Richard Eden published an English translation.
[‡]Recorded in Richard Haklyut's book *The Principal Navigations, Voyages, Traffiques and Discoveries of the English Nation*, published in 1598.

that the deadly effect of curare was not on the heart but on the muscles of the lungs. However, it was at Leiden in 1775 that further research on curare showed that breathing the fumes given off when it was heated did not kill as had been reported by early explorers, nor did eating the flesh of animals that had been killed with it have an effect. However, merely piercing the skin with a lance tipped with curare quickly led to death.

The effects of curare were demonstrated to members of the Royal Society of London in 1812 when Sir Benjamin Brodie injected a guinea pig with it, whereupon it stopped breathing although its heart continued to beat. He then showed that the animal could be kept alive by artificial respiration, using bellows to pump air into its lungs. Brodie continued to do this for 20 minutes until the animal recovered, and it was reputed to live for more than year after the experience.

Meanwhile, in France, another researcher was working on curare, seemingly unaware of any previous research. In the 1840s, Claude Bernard (lived 1813–1878) worked out how curare killed its victims. His first experiment was on a rabbit, which he impaled with an arrow tipped with the poison. He noted that within five minutes the rabbit fell over while its heart continued to beat but it did not respond when he pinched it. He also experimented on feeding a rabbit with curare and noted it had no effect. He published his findings in 1856.

The earliest report of using curare medically was made by Sir Robert Hermann Schomburgk (lived 1804–1865), a German explorer who was employed by the British to carry out geographical and botanical studies in South America and the West Indies. He went down with malaria, and when his supply of quinine ran out, he tried taking small doses of curare for the fever from which he was suffering. However, it had no effect and he reported as such in 1841. Nevertheless, in the latter part of the 1800s, doctors began to prescribe curare to treat tetanus, epilepsy, and muscle spasms, although it could be only of minor benefit.

Others continued to experiment with curare, generally with negative results except when it came to understanding how the body's nerves operated. How did messages pass from the brain to vital organs like the muscles? For many years, a controversy raged as to whether this relied entirely on electrical impulses along nerve fibres or whether something else was involved. The puzzle was that there was a gap between the end of the nerve fibre and that of the organ which it was stimulating. How did a signal then pass from one to the other? The answer was the formation and release of the chemical messenger **acetylcholine** and it was this which carried the message across the gap. Nerve endings maintained a supply of this chemical ready for use. The two men eventually to prove this were the Austrian Otto Loewi and the Briton Henry Dale, and they shared the Nobel Prize in Chemistry in 1936 in recognition of their work.

Curare kills by blocking acetylcholine receptors and so prevents acetylcholine from operating as it should, and especially with regard to the vital muscles of the body, like those of the lungs. With these sites blocked, the nerve cannot respond to a signal. A treatment for curare poisoning is **physostigmine** (aka **eserine**), which inhibits the enzyme acetylcholinesterase.

Research into curare was performed mainly by Harold King (lived 1887–1956) of the Institute for Medical Research, at Mill Hill in north London. In the 1930s and 1940s, he managed to separate 1.18 g of the active agent, which he named tubocurarine chloride. Curare, as such, was a mixture of chemicals, although tubocurarine predominated and this became the official name for the drug.[§] King showed its chemical composition was $C_{38}H_{44}Cl_2N_2O_6$, and by 1935, he had deduced its structure although, in fact, he had analysed it slightly wrongly as having 37 carbon atoms not 38. In 1970, its correct structure was finally determined and King shown to be almost correct.

Another chemical that is present in curare extracts is suxamethonium, and this is used medically to relax muscles in the throat to make it easier to insert a flexible plastic tube into the windpipe to keep this open, or through which to deliver certain drugs.

11.3 CURARE BECOMES PART OF SURGERY

The American Richard Gill spent several years in the 1920s living with the natives of Ecuador, and they taught him how to prepare curare. On his return to the USA, he had with him around

[§]It was given this name by Harold King in 1935, and he chose it because the curare he obtained came in a tube.

12 kilos of the toxin and hoped to get drug companies interested in it, but with little success. Then he wrote the book *White Water and Black Magic*, and this did arouse interest at two drug companies whose researchers contacted him. They could see its potential as a muscle relaxant.

As more curare became available, other companies began to experiment with it, and at Wellcome, a UK drug company,[¶] the director of clinical research, Dr Frederick Prescott, even had himself injected with it to demonstrate that it could be used safely.

The first person to use curare as a muscle relaxant during surgery was a Dr Abram Bennett of the Clarkson Memorial Hospital of Omaha, Nebraska, in 1940. News of what he was doing reached Harold Griffith (lived 1894–1985) at the Queen Elizabeth Hospital of Montreal, and on 23 January 1942, he used it for the first time and was so impressed he quickly published his findings. He is now regarded as one of the leading individuals who made anaesthesia a safe part of surgery.

However, curare did not come into general use until the 1950s, and then it had a profound effect. By relaxing muscles, it allowed surgeons to open the stomach wall and operate deep within the body. Thanks to curare, it was possible to reduce the amount of anaesthetic being administered and this in itself was beneficial. The story of how curare came to play such a role is admirably told in Stanley Feldman's book *Poison Arrows*, published in 2005, in which he describes the way this chemical was analysed and how its effects on the body were studied scientifically for more than a hundred years.

Curare was eventually superseded by safer relaxants, and today the most widely used muscle relaxant is rocuronium bromide. Provided the patient is supplied with artificial respiration, then this can be used and, like curare, it will eventually be eliminated from the body with no harm done.

11.4 MULTIPLE MURDERS, NEW JERSEY, 1965–1975

Mario Enrique Jascalevich originally came from Argentina, and he used curare to kill the patients of another surgeon in the

---

[¶]Wellcome is no longer a drug manufacturer and its pharmaceutical division is now part of GlaxoSmithKline.

hospital towards whom he felt inferior. Jascalevich was finally brought to justice many years later but was found not guilty after a lengthy trial. He had employed a lawyer who was able to attack the forensic evidence by showing that his client was being accused on the basis of findings which could not be supported by the chemical research methods available at the time. Soon after his trial ended, Jascalevich flew back to his home country, where he died in 1984, aged 57. It had been said that he never paid his lawyer's fees.

Jascalevich was born in Buenos Aires in August 1927 and trained as a doctor in Argentina before moving to the United States in 1955, where he did an internship at St Mary's General Hospital in Passaic, New Jersey. He then set up his own practice, and in 1962 he became a surgeon at Riverdell Hospital in Oradell, New Jersey.

Jascalevich was not only a competent surgeon but also an active researcher. He published his findings in reputable medical journals like *Surgery*, where 12 original papers of his appeared in the years from 1960 to 1984. In the first of these, he reported a new kind of suction cone for draining sinuses. One paper in particular, published in 1967, reported a new type of stapler for gastric operations, and this became widely used for several years and was named after him.

So, why did this successful surgeon and innovative researcher resort to murder? Although we can never really know the answer to that, we do get a hint of the kind of man he was when we realise that he did not murder his own patients, but those of other surgeons. His first murder was probably committed in 1965, and the victim was a man of 73 who was waiting for a hernia repair. When Jascalevich examined him, he decided that because the man was suffering from a heart condition, albeit mild, he should not be operated on that day and the man was put on an intravenous drip. He died soon afterwards.

His next victim was a four-year-old girl who had been admitted to hospital with acute appendicitis. She had been successfully operated on by fellow surgeon Stanley Harris. The girl came round from the anaesthetic and was on the road to recovery when she suddenly died and for no apparent reason. Other unexpected deaths followed, so much so that Harris and a colleague, Dr Allan Lans, began to investigate what was going on.

All the victims had an intravenous drip at the time of death and all died early in the morning when Jascalevich was in the ward. Strangely, none of his own patients died this way, only those of other surgeons.

Harris was sufficiently suspicious of Jascalevich that he took the unprecedented action of opening Jascalevich's locker, where he found vials of curare, some of which had been opened and others had been half used. This was enough to alert the director of the hospital, who obtained a warrant to conduct a formal and forensic search of the locker, and that commenced on 1 November 1966, when the presence of curare was officially noted.

Jascalevich was asked what was going on. He said he was using the drug in some experiments on dogs and showed the investigators his research laboratory where he had been carrying out these experiments. Indeed, when the contents of his locker were more closely examined, there were traces of dog blood and dog hairs present, although he was never able to explain how the dogs he said he was experimenting on had been obtained. Some of the curare had been purchased by Jascalevich for these experiments, although some of the vials in the locker appeared to have been taken from the hospital's own supply.

Faced with the accusations made against him, Jascalevich resigned from Riverdell Hospital in 1967 and found employment elsewhere, and thereafter the sequence of sudden and unexplained deaths at the hospital came to an end. Things did not rest there, however.

There were those who were convinced that Jascalevich had been guilty of multiple murders, and in 1975 they sent a letter to the *New York Times* saying that the sequence of strange deaths at Riverdell should be looked into, and how these might be linked to a certain surgeon. A reporter, Myron Farber, was assigned to the case and began to investigate. He was given access to the hospital files and the names of those who had suddenly and inexplicably died, so that he could interview their families. He wrote three long articles for the newspaper in which he referred to a "Dr X" and claimed that 13 patients at Riverdell Hospital had died in unexplained circumstances and that Dr X was to blame.

The Bergen County Prosecutor reopened the case of the curious deaths at the hospital after the New York Deputy Medical

Examiner, Dr Michael Baden, swore an affidavit stating that at least 20 deaths at Riverdell in 1966 were misreported on death certificates. He too reviewed the case notes and was of the opinion that many of the deaths could not be explained. He said that they had been caused by the respiratory depressant curare and that some of the bodies of the deceased should be forensically examined, if that were possible.

Five of their next-of-kin agreed to exhumations and tissue samples were taken for analysis, and several of these tested positive for curare – or so it was claimed. In April 1976, Jascalevich was suspended but this action was declared null and void by Superior Court Judge Frederick Kentz Jr. However, Jascalevich voluntarily agreed to stop operating forthwith, but in any case, he was soon to be charged with murder.

A Superior Court judge agreed that five bodies should be exhumed: those of Emma Arzt (aged 70), Carl Rohrbeck (aged 73), Frank Biggs (59), Margaret Henderson (aged 26) and 4-year-old Nancy Savino. All had been admitted to Riverdell for routine surgery and had died unexpectedly while apparently recovering from successful operations.

Jascalevich's trial began on 28 February 1978 and was to last a record 34 weeks. Sometimes there was a surprising comment from witnesses suggesting that news of Jascalevich's behaviour was already well known on the medical circuit. On 4 May 1978, one of the doctors, Dr Edwin Friedman, who allocated patients to the surgeons at the Oradell, New Jersey hospital, was in the witness stand and he was asked why he continued to refer patients to Jascalevich: "The safest thing was to send patients to him because nothing happened to those patients." Friedman believed that Jascalevich objected to new surgeons who joined the hospital staff because he wanted to have the field to himself.

In the witness box the following week was one of the young surgeons at Riverdell, Robert Livingston. He said he particularly remembered the first operation he did there, which involved a routine abdominal exploratory operation of Margaret Henderson, who had been admitted to hospital with severe abdominal pains. He examined her and said he could feel a pelvic mass, but Jascalevich disagreed and suggested a barium enema and X-ray, which did indeed show a large lump in her lower abdomen. Livingston then performed an operation but the mass was due to an enlarged uterus and ovarian cyst which was removed. However, she died the following morning quite unexpectedly. Somewhat oddly, Jascalevich rang Livingston at his home and said he should get back to the hospital immediately. Jascalevich said that he thought Mrs Henderson seemed dehydrated and so he had put her on an intravenous drip. That in itself was rather strange because it was not something that a surgeon would normally do himself. That was a nurse's responsibility.

The detection of curare in specimens taken from the corpses of his supposed victims was to be a key part in Jascalevich's trial. Naturally, such evidence was to be challenged by the defence and in several ways. Could curare survive for 10 years in a body that had been embalmed? How might the embalming fluid affect the curare? Could the curare really have transformed into something else that might be identified as deriving from the drug? How was curare to be identified in small amounts? The defence concentrated on these issues, and even if curare was proved to be present in a body, did that necessarily mean that Jascalevich had injected it?

The prosecution needed to answer these points and they drew on the most advanced methods of chemical analysis at the time. One witness, David Beggs, had analysed samples by mass spectrometry (MS), and he could definitely identify curare in the remains of one victim, Nancy Savino, and possibly in two other victims. Beggs was from the Hewlett Packard Corporation, who were the main manufacturers of analytical equipment.

Another witness was Dr Leo Cortivo who said that he had found curare in the tissue remains of three of the corpses, and he used high performance liquid chromatography (HPLC) and radioimmunoassay (RIA) to determine this. He also had found curare in the vials in Jascalevich's locker. (The judge decided that because no curare had been found in the bodies of Emma Arzt and Margaret Henderson, the case against Jascalevich with respect to them should be dropped.)

The principal witness for the defence was Abraham Stolman, Chief Toxicologist for the State of Connecticut, and he pointed out that the various methods of analysis used by the prosecution witnesses were not sensitive enough for their finding of curare to be reliable.

The defence also had an expert scientific witness, Frederick Rieders, a toxicologist from Philadelphia. He said only the MS results would prove curare was present and his research showed that, even were it to be present, if it had been in contact with embalming fluid it would decompose and be undetectable within a few weeks. Clearly, samples taken from bodies that had been interred in 1966 would have had to have a great deal of curare in them if any was to survive to 1976. And yet Rieders had to admit that he too had detected curare in the liver of Nancy Savino, but only in the liver. The defence suggested that accidental contamination of the sample with curare had occurred when the child's liver had reached the laboratory.

When, in October 1978, the jury retired to consider their verdict, it was expected that it would take them a long time to digest the analytical results and compare the findings of both sets of experts. In fact, they returned to the court room two hours later and delivered a unanimous verdict of not guilty. And the jury went even further, to the extent of writing to Dr Albano, New Jersey State Medical Examiner, to say not only was Jascalevich not guilty of the charges made against him but that they were convinced he was entirely innocent and that he should be allowed to continue his medical practice.

It seemed that Jascalevich had got away with murder. However, his licence to practise medicine was revoked by the New Jersey Medical Licensing Board by a vote of 11 to 0. At the time, Jascalevich was practicing in the Bronx under a New York State licence, and this meant he could no longer work as a surgeon. Jascalevich returned to Argentina and bought a house in Mar del Plata, Argentina's second largest city, and there he died of a cerebral haemorrhage, aged 57. How many patients he murdered at Riverdell Hospital is not known for certain. Admissions to this private hospital decreased significantly, and it closed and was demolished in 1984, although by then it had changed its name to Northern Community Hospital. Even so, only a third of its beds were ever occupied.

A spin-off of the Jascalevich case was still to be played out. When the defence lawyer, Mr Brown, had tried to get the *New York Times* reporter's notes by means of a subpoena, Farber refused to give them up because he had promised his informants that he would protect their identity. The result was that he was sent to jail for 40 days for contempt of court and the *New York Times* was fined $286 000 (in January 1982, both Farber and the newspaper were pardoned and half of the fine money was returned).

CHAPTER 12

Aconitine and Wimbledon

A word in **bold** *indicates that further information can be found in the Glossary. Only the first time the word appears in a chapter will it be so indicated.*

The flowering plant known as monkshood was so called because of the shape of its flowers. Its botanical name is *Aconitum napellus*, and it is popularly referred to as aconite. This is the most poisonous of all plants – its roots especially so – because it produces the chemical **aconitine**, sometimes referred to as the Queen of Poisons (Figure 12.1). It has other names, such as wolfsbane, because the paste made from it was used to tip arrows, and when one of these hit a wolf, then the animal was sure to die. The plant itself is thought to be a native of India, although it came to Europe many centuries ago and may even have been known to the ancient Greeks.

Aconitine is dangerous because it can be absorbed through the skin; just touching the leaves may lead to numbness in the fingers. Sometimes warnings have to be issued in Southern China when small amounts of aconite leaves have entered the food chain and people have consumed them in soups and salads and been made ill.

More Molecules of Murder
By John Emsley

Published by the Royal Society of Chemistry, www.rsc.org

Figure 12.1 *Aconitum napellus*, the Queen of Poisons. © Manfred Ruckszio/ Shutterstock.

12.1 ACONITINE AND ITS TOXICITY

Aconitine is an **alkaloid** like **strychnine** and digitalis, and it is made by the plant, presumably to deter anything which might wish to eat it. Every part of the plant contains it. When it is ingested, it is quickly absorbed into the body with painful and fatal consequences.

So how toxic is aconitine? Its **LD_{50}** is assessed to be 30 micrograms per kilogram so that for an average 70 kg person a dose of 2 mg might well kill and a dose of 4 mg would almost certainly do so. This means that ingesting less than a gram of the plant's leaves could be fatal.

This is what happened to a Canadian actor, 25-year-old Andre Noble, in 2004. He went to stay with his aunt at her cabin on Fair Island, Newfoundland. He was a rising star of TV and films and appeared to have a successful career in front of him. He even co-founded a theatre group called The Young Company. One day, while hiking with his aunt, he picked a few leaves off a monkshood plant and ate them. He quickly became ill and they rushed back to the cabin. An ambulance was sent for, but he died on his way to hospital.

The symptoms that aconitine produces are acute stomach pain, excessive vomiting, diarrhoea, irregular heartbeat, and then death. Aconitine interferes with the sodium channels of the central nervous system, keeping them locked open and thereby preventing them from conveying key information along the central nervous system. The body requires these channels to open and close in order for messages to flow; leaving them open breaks this line of essential communication.

There is no known antidote for aconitine. When someone has been poisoned by it, their treatment has to concentrate on keeping the heart beating regularly, and this involves giving drugs such as amiodarone or flecainide. Activated charcoal can be given to absorb the poison in the stomach but this must be done quickly, and within 15 to 20 minutes, if it is to be effective. The half-life of aconitine in the human body is six to eight hours, although it may be longer if the liver and kidneys are impaired because these are where the poison migrates to, there awaiting elimination *via* the urine.

In addition to accidental poisoning, aconitine has also been used by those wishing to commit suicide or to murder. In the Indian Mutiny of 1857, an attempt was made by some cooks to poison officers of the British East India Company by putting aconitine into their curry. One of the officers, John Nicholson, learned of the plot and confronted them. He demanded that they eat some of the food they had prepared. When they refused to do so, he had some force-fed to a monkey which quickly died. He then had the chefs executed by hanging.

Aconitine became notorious in 1881 as a result of a famous murder, more details of which are given below. This is probably why it features as the poison of choice in Oscar Wilde's story *Lord Savile's Crime*, written in 1887. Lord Savile has his fortune told and is informed that he will only marry after he has murdered someone. He attempts to poison his aunt, Clementina, and he gives her a capsule containing aconitine, saying it will cure her heartburn. Then, while visiting Venice, he learns that she has died and so he returns to London to reclaim the inheritance she bequeathed him in the form of some property. Now, he assumes he can marry as was foretold. However, his wife-to-be discovers the deadly pill, which his aunt had failed to take. The wedding has to be postponed, and so Saville has to find another victim to murder.

Aconitine also featured in James Joyce's stream-of-conscience novel *Ulysses*, which records the thoughts of one man, Leopold Bloom, over the period of one day. Aconitine, as monkshood, is mentioned because it is the poison that Bloom's father used to commit suicide.

12.2 MEDICAL USES OF ACONITINE

Today, it seems strange that a chemical as deadly as aconitine was ever used by doctors to treat patients, but as we have seen, quite a few poisons of plant origin have ended up in medicines at one time or another. The dried root of aconite can contain as much as 0.5% aconitine and was once available from pharmacists. Aconitine in a pure form can be extracted from aconite by means of solvents like chloroform or alcohol and, by subsequently evaporating the solution, a white crystalline powder is obtained which is mainly aconitine.

Like other toxic plants mentioned in this book, aconitine was used by doctors as a pain-reliever, but only in tiny amounts. However, the dose between safe and toxic is narrow and, not surprisingly, it was a risky treatment. It was also given to reduce fevers. The use of aconitine by doctors was more prominent in France than in the UK, mainly as an ointment for skin complaints or to treat the ears, but was never prescribed for internal use. In contact with the skin, a salve containing aconitine generates a tingling sensation followed by numbness and, for this reason, it was used as a localised treatment for rheumatism, neuralgia, and sciatica.

Aconite roots have been a part of traditional Chinese medicine but this is only after soaking in water and boiling them, which converts the aconitine they contain into non-toxic molecules. Even so, there is always a risk of poisoning.

There have been relatively few murders using aconitine, even though it is one of the deadliest poisons known. One reason is that symptoms of poisoning appear almost instantly, such as a burning sensation in the mouth. Two would-be murderers dealt with this in different ways: the first murderer gave his victim the poison in a capsule, which he was induced to swallow; the second murderer put the poison in a curry dish whose hotness would mask the effect, just as the Indian cooks had planned to do in 1847. The first murder occurred in 1881, the second in 2009, and both happened in London.

12.3 MURDER, WIMBLEDON, 1881

George Henry Lamson was born in 1852 in America, and he trained as a doctor. He then came to Europe and was a volunteer surgeon in the Franco–Prussian war of 1870–71. He worked with the French Ambulance Corps during the siege of Paris, for which he was awarded the Legion d'Honneur medal. He played a similar role in the Romanian War of Independence of 1877 and again he was awarded a medal. He came to live in Great Britain the following year, and by then he was addicted to morphine, a drug he had often used when dealing with badly wounded soldiers.

In 1878, 26-year-old Lamson took up residence in Bournemouth and established himself as a GP. He married Kate John, the daughter of a wealthy Manchester merchant, William John, and thereby acquired control of her finances, which he needed because his addiction was proving very expensive. He began to leave bills unpaid or to pay with cheques that bounced. He was also in arrears with his rent. By 1880, Lamson was in dire financial straits, owing money to friends and acquaintances. (He even pawned his surgical instruments for £5 at a London pawnbroker.) But he knew there was a source of money that would eventually come his and Kate's way. It was in the possession of his brother-in-law, who was an invalid and unmarried. All it needed was for him to die.

Percy Malcolm John was 19 years old and suffered from a pronounced curvature of the spine which meant he could only move about using a wheelchair or if someone carried him. He was a long-time resident at Blenheim House School in Wimbledon. Percy's father had died and left his money to his wife. She died in 1869, and the wealth was then divided among her three children. Needless to say, the portion inherited by Kate had soon been used to clear Lamson's debts and feed his addiction. Percy's inheritance was invested in Government Bonds to the value of around £3000, which produced an income for him of £109 per year (the equivalent amount in today's money would be £300 000 and £10 900 respectively). If Percy were to die, this would then be divided between his two sisters. Kate would thus inherit £1500 (equivalent to £150 000 today), and this would again fall under the control of Lamson, although legally it was Kate's according to the 1870 Married Woman's Property Act.

In December 1881, Lamson planned a visit to Florence to see his father and decided to call at Blenheim House in Wimbledon while passing through London. He had put some aconitine into a capsule and his aim was to persuade Percy to swallow this, believing it contained only sugar. In a letter to Percy, he said that he would call on him on Saturday the 3rd but their meeting would have to be brief because he was leaving for Paris that evening on his way to Florence.

Lamson arrived at the school as planned and he brought with him a Dundee cake as a treat for Percy. It was an expensive cake which he had bought at a confectioner's shop down Oxford Street in London. All of those present in the room with Percy and Lamson had a slice of this special treat, including the headmaster, Mr Bedbrook, who had come to meet Lamson. He offered the doctor a glass of sherry and even Percy had some, although he said he would like some sugar to sweeten it. The headmaster sent for the matron and she brought some caster sugar from the kitchen.

Then Lamson spoke of a new kind of capsule which he had brought with him and he would like to show the headmaster how easy it was to swallow this new method of taking medicine. Lamson showed him one and then put some of the caster sugar into it, gave it a shake and handed it to Percy, who swallowed it easily. The headmaster was impressed and Lamson even left some empty capsules for the school to use if necessary. It was then time for Lamson to catch his train back into London, and as he was leaving, he said to the headmaster how much worse Percy's condition was and that he didn't think he would last much longer.

Soon after Lamson had left, Percy complained of heartburn and was carried up to bed by the student he shared his room with, which was the usual procedure. The student then left him for a couple of hours, but when he too went up to bed, he found Percy in the bathroom vomiting. (Some of this vomit, which was in the bath, was collected the following day, after Percy had died.)

An hour later, Percy's suffering was far worse, and he was displaying all the symptoms of aconitine poisoning and writhing in pain to such an extent that the teachers had to hold him down. He died at 11:30 pm. It was clear that Percy had been poisoned – by the cake or by the capsule – and that his uncle had done it. Lamson was now in Paris and there he learned from a London newspaper that he was responsible for the boy's death. He returned to London pleading his innocence but was immediately arrested and charged with Percy's murder.

Lamson had chosen aconitine because he thought its presence in the body could not be proved because so little was needed to kill the victim. Indeed, that might have been the case a few years earlier but things had moved on, as Lamson was to discover during his trial. This began at the Old Bailey in February 1882, two months after the murder. Lamson stood in the dock dressed in a manner befitting his calling and appeared to be a respectable doctor. He made copious notes during his trial. However, the forensic evidence proved conclusively that Percy had been poisoned with aconitine.

On Saturday 11 March, the court at the Old Bailey heard the findings of Dr Thomas Stevenson, a specialist in poisons based at London University. He had been asked to analyse the various samples collected after the death of Percy, which included vomit, urine, his stomach contents, as well as various medicaments that had been in Lamson's possession. Stevenson extracted the samples with solvents, which he then evaporated off and tasted the residue. He had also obtained a sample of the drug for comparison purposes. The burning sensation which aconitine produced when put on his tongue ("like a hot iron") was so strong that it was instantly identifiable. This effect was noted when extracts from all Percy's bodily fluids were tested. But, just to confirm its deadly nature, some of the material extracted was injected into mice, which invariably died within a few minutes. The remains of the Dundee cake tested negative.

Stevenson also analysed the contents of the ampules found in Lamson's possession, and while most of them tested negative, two were found to contain pure aconitine. Stevenson told the jury that the effect of that on his tongue lasted for seven and a half hours. One of the pills left at the school by Lamson also contained traces of white powder, and this proved also to be aconitine.

That Lamson had possessed aconitine was confirmed by a pharmacist who worked near the hotel where Lamson had stayed while in London. It was revealed that on 1 December Lamson had bought two grains (about 130 mg), for which he paid two shillings and sixpence (equivalent to £12 today). That, together with the forensic evidence, persuaded the jury that Lamson was guilty, and they took less than half an hour to reach their decision. (His defence maintained that there was no absolute proof that Percy had been poisoned with aconitine, or that Lamson had given it to him.)

His execution was set for 4 April but was deferred when the US President, Chester Arthur, requested a delay to enable evidence of Lamson's insanity, which was claimed to be hereditary, to be sent to London. This showed that his grandmother and other relatives had been detained in the Bloomingdale Asylum for the Insane at various times. However, this did not convince the Home Secretary that Lamson was insane, and he was duly hanged at Wandsworth Prison on the morning of 28 April 1882, after a last meal of scrambled eggs and toast with coffee.

Lamson may even have tried to poison Percy the previous summer when the boy was on holiday at Shanklin on the Isle of Wight. On that occasion, he also gave him a quinine pill, a drug famous for its bitterness which he no doubt hoped would disguise the aconitine, but the pill contained only a small dose of aconitine and not enough to kill Percy. Soon after Lamson's visit, Percy fell very ill and said he felt paralysed all over – but he eventually recovered. At the time, Lamson was staying with his mother who lived at Ventnor, a few miles from where Percy was staying. A chemist on the island later confirmed that he had sold Lamson some quinine, **atropine**, and a grain of aconitine.

12.4 MURDER, SOUTHALL, WEST LONDON, 2009

In this case, the murderer tried three times to achieve her goal: to punish and poison her former lover and his new girlfriend.

Lakhvinder "Lucky" Cheema was 39 and lived in Southall, west London. For 16 years, he had had a secret affair with 45-year-old Lakhvir Kaur Singh, whose 57-year-old husband was being treated for cancer and by whom she had had three children. The affair started when Cheema came to live with the Singhs after his own marriage had failed. Twice, their love-making resulted in Singh becoming pregnant, although both times she had an abortion. When Cheema moved to a houses of his own and took on lodgers, Singh visited him regularly, ostensibly to act as a cleaner and to do the laundry, and so things continued, but it was not to last.

In 2008, Cheema ended the affair under pressure from his own parents, and so he sought someone he could marry and with whom he could have a family. He began to date 22-year-old Ms Choongh, an illegal immigrant, and in November 2008, they got engaged and planned to marry the following year on Valentine's Day, February 14. Singh was distraught on hearing of this and begged Cheema to break off the engagement but he refused.

One evening, she even discovered the lovers in bed together. It was just too much. If Singh could not have Cheema then no one could. She was furious and threatened to burn the house down, and she may even have tried to poison him then, although with what was never established, but clearly it was not successful. It is impossible to purchase chemicals in the UK which are dangerously poisonous, so she may have had recourse to using an insecticide or pesticide, neither of which would be capable of killing a human.

Singh decided that if she was going to murder Cheema it would have to be with something much deadlier, and that kind of compound could only be obtained in India. She flew out there and stayed for three weeks, during which time she purchased some aconitine. On her return, it seems she attempted to poison him again but used only a tiny amount. He was made ill enough to need hospital treatment for a week, although the doctors were unable to say what had caused his upset stomach and vomiting. Meanwhile, Singh visited him every day in hospital, and it was then that fiancée Choongh realised that she had had a relationship with Cheema and warned her to stay away from him.

On 27 January 2009, Singh went round to Cheema's home, where she let herself in with the key she still had and mixed some of the poison with a curry in the fridge. Cheema loved curry and he had prepared this the previous evening and now planned to share it with Choongh. Cheema would not have hesitated to eat a dish he had prepared himself and there would be no reason for him to assume it had been tampered with. However, a student lodger in the house had seen Singh in the kitchen and noted that she had taken the curry from the fridge, but thought nothing of it. However, the student was not around when Cheema came home from work, and so he and Ms Choongh unsuspectingly ate the curry dish for their evening meal.

Soon after they had eaten, they began to feel the effects of the aconitine, first in the form of pins and needles around their mouths. Choongh thought she would feel better if she took a shower, which she did but to no effect. Soon they were vomiting and everything seemed to get darker as their sight was affected. Cheema rang for an ambulance, telling the operator that he and his girlfriend had eaten food that had been poisoned. When asked how he knew, he replied that his ex-girlfriend had done it. Rather than wait for an ambulance, Cheema persuaded his nephew who lived with them to take them to a nearby hospital in his car.

At the hospital, and still vomiting, Cheema told doctors that he suspected his former girlfriend had put something in the food he had eaten. By now his heart was racing, his vomiting was getting worse and he began to have convulsions. Two hours after he had arrived at the hospital, he died. Ms Choongh's heart also began to behave erratically and she was placed in a medically induced coma while attempts were made to stabilise her condition, and it worked. A week later, she was well again.

Meanwhile, Cheema's house was evacuated and searched for whatever it was that had poisoned the couple, but nothing was found. Next, the police approached Singh and searched her, finding a plastic bag in her handbag which contained a brown powder. This she said was to treat a rash on her neck. But what was it? Aconitine was not a poison that British forensic chemists were familiar with, but they could show that whatever the brown powder was, it was the same chemical that was in the suspect curry and the vomit.

The forensic investigation was more challenging than normal, and it fell to Denise Stanworth at the Laboratory of the

Government Chemist (LGC Limited) in Abingdon to say what had poisoned the couple. She thought it might be an alkaloid from *Aconitum ferox*, which grows in the Himalayas. To confirm this, she consulted the UK's Royal Botanic Gardens in Kew to obtain a reference sample to compare with the powder found in Singh's possession, and this confirmed it did indeed come from *Aconitum ferox*. Cheema's death was definitely due to aconitine.

Singh's trial was held at the Old Bailey in January 2010, and she was found guilty of murdering Cheema and given a life sentence with a minimum term of 23 years. She was also found guilty of causing Choongh grievous bodily harm.

CHAPTER 13

Cantharidin and Spanish Fly

A word in **bold** *indicates that further information can be found in the Glossary. Only the first time the word appears in a chapter will it be so indicated.*

Spanish fly is the popular name for the powder made from a beetle which, throughout history, has supposedly had aphrodisiac properties. In fact, Spanish fly is a misnomer, as it is neither Spanish nor does it come from a fly. Nevertheless, it was seen as something that turned men on, the most obvious effect of which was a long-lasting erection, and it was supposed to increase the desire for sex in both men and women.

Cantharidin is present in beetles of the *Meloidae* group of insects and there are more than 200 species of them, many being brightly coloured. They are more commonly known as blister beetles because all of them can produce a toxic agent, which they excrete from glands in their joints. In parts of the USA, they have been responsible for the deaths of horses which have eaten hay that has been infested by them. Cantharidin is also highly toxic to cattle, sheep, dogs, cats, rats – and humans.

A solution of beetle extract was used by doctors and was called cantharides. Eventually, the active chemical was identified by Samuel Coffey in 1923. Cantharidin could be prescribed with more confidence because the exact dosage could be determined,

More Molecules of Murder
By John Emsley

Published by the Royal Society of Chemistry, www.rsc.org

whereas with Spanish fly the amount of active agent could vary considerably depending which type of beetle had been used. In fact, all these insects were best avoided because cantharidin is not only a deadly poison but it kills in a most vicious way, and there is no known antidote.

Cantharidin, as Spanish fly, featured in the TV show *Whitechapel*, broadcast in 2012, and in Roald Dahl's first adult novel, *My Uncle Oswald*, which is set in 1919. It tells the story of Oswald Cornelius, who had discovered the benefits of the Sudanese blister beetle and used it to steal the sperm of leading scientists like Albert Einstein and artists like Claude Monet, and then sold it to rich women. The most recent reference to Spanish Fly was in Mike Batt's popular song *My Aphrodisiac is You*, which Katie Melua released in November 2003. One verse of her song has the lines:

> "Don't smoke no grass, or opium from old Hong Kong,
> That hubble-bubble just makes me see you double all night long.
> Don't waste my time with Spanish fly and roots to chew,
> They cause me trouble, because my aphrodisiac is you."

13.1 CANTHARIDIN

Cantharidin can be extracted by organic solvents – it is barely soluble in water – and as the solvent evaporates, it leaves behind colourless, odourless crystals. Cantharidin was first extracted from crushed beetles in 1810 by a French chemist called Pierre Jean Robiquet (1780–1840).[†] Although chemists knew the chemical formula of cantharidin, and eventually its molecular structure, it was not until 1951 that it was successfully made in the lab, and that was achieved by an American chemist, Gilbert Stork. It is still not known how blister beetles synthesise this molecule, which consists of three interconnected rings.

Cantharidin plays a vital part in the mating of certain insects such as the bright green blister beetle, whose scientific name is *Lytta vesicatoria,* and this was the main source of Spanish fly (Figure 13.1). This creature has a curious reproductive cycle in

[†]He was also the first to extract **caffeine** in 1821 and codeine in 1832.

Figure 13.1 *Cantharis* or *Lytta vesicatoria*, Spanish fly or blister beetle, vintage engraving. © Morphart Creation/Shutterstock.

that it lays its eggs near the nest of a ground-nesting bee, and when the eggs hatch, the larvae crawl into the bee's nest and take over.

The adult male beetle excretes cantharidin from its joints and makes a sticky ball from it. This is presented to the female he wants to mate with and she sniffs it. If she likes what she smells, she will offer herself to the male and, having mated, will take his cantharidin to place with her eggs to protect them against predators. A typical beetle generates around 0.5 mg of cantharidin.

Some beetles make much more cantharidin than others, the range being from 0.2 mg to 4.8 mg, the latter being the level in *Epicauta immaculata* (aka big gray). Around 1500 species of beetle produce this molecule, and they can have a very distinctive colouring. If such a beetle lands on your skin, you should not try to kill it or even brush if off as it may react by releasing cantharidin. Try and remove it gently, for example by blowing on it.

Cantharidin is still used in animal husbandry to increase fertility, just as it did for humans for thousands of years. However, it does not work on the male body in the way that Viagra does, although it does have the same effect of producing a prolonged erection in men. It can also stimulate women into wanting sex through its effects on the vagina. It results in a condition referred to as pelvic vascular congestion, which results in the accumulation of blood in those parts of the body we most associate with the sexual act. Nor are humans the only ones to have made use of it. The great bustard is a large bird which lives in open grasslands across southern Europe, and the male will deliberately seek out and eat blister beetles during the mating season.

13.2 CANTHARIDIN DOWN THE CENTURIES

Hippocrates, the Greek physician who lived 450–370 BC, said that a powder made from dried blister beetles was a treatment for dropsy. At the time of the Roman empire, another Greek physician and pharmacologist, Pedanius Dioscorides, who lived in the first century AD, wrote *Material Medica*, and this too has a section on cantharidin and its use for treating various ailments. Whether it worked or not, these ancient medical men clearly knew of cantharides but were advocating the use of one of the deadliest of natural poisons.

In China, the dried body of the Chinese blister beetle has been used in folk medicine for more than 2000 years. It was a traditional treatment for all skin conditions, including ulcers, as well as a way to deal with bowel complaints such as worms and even piles. For men, it was regarded as an aphrodisiac, while for women, it was seen as something that would cause an abortion, and had been so used.

Those who bought Spanish fly with a view to using it to increase sexual potency needed to know just how much, or how little, was needed. Nor need it be taken directly, as one group of French soldiers, based in North Africa, discovered in 1893. The frogs they were partial to eating had, in their turn, eaten *Cantharis* beetles and this caused the men to suffer prolonged erections, as their regiment doctor discovered. Given the right amount, the victim would not suffer the blistering effects within their body yet there would be enough to irritate the bladder and urethra, causing them to swell; and while this might make passing urine somewhat difficult, the irritation it caused might also result in more blood entering the penis or vagina, making the owner think about sex.

The empress Livia, wife of the great Augustus Caesar who was emperor of Rome for 40 years from 27 BC to 14 AD, is reputed to have fed Spanish fly to some of the men she invited to dinner, in the hope that this would encourage them to engage in illicit sexual behaviour and so compromise them. Clearly, knowledge of this remarkable compound persisted down the ages. In 1572, Ambroise Paré reported that Spanish fly had caused a man to suffer an erect penis for several days. The infamous French poisoner of the 17th century, knows as La Voisine,[‡] sold a love charm which included Spanish fly for women to give to their partners.

In the 18th century, it was made notorious by the Marquis de Sade, who clearly knew the effects it could produce. In 1772, he persuaded a group of prostitutes from Marseilles to take part in an orgy in which Spanish fly would be used to boost their performances. However, he gave them more than was necessary and they became very ill. De Sade panicked and fled to Italy with his sister-in-law, Anne Prospère de Launay, with whom he was having an affair. Her aristocratic mother got a warrant for his arrest signed by the king and he was duly imprisoned.[§]

Spanish fly was also available in Great Britain and sometimes deliberately misused, not for one's own benefit but to generate an effect on a desired individual, as the trial of Thomas Neeve at the Old Bailey on 7 December 1687 showed. He lived in Westminster and was accused of giving Ann Oliver a drink to which he had added some cantharides, which made her very ill, although she survived. He was found not guilty.

Just how big a trade in cantharides there was in London was revealed at a trial held in 1825. Elizabeth Smith was accused of stealing cantharides from her employer Samuel Foulger on 23 July that year. His shop was in Ratcliffe Highway in the East End of London, and Foulger was beginning to suspect that Ms Smith was up to no good, so he deliberately got up early one morning, and with his son, they hid and watched what she did. She took some cantharides from a cask, put it in a paper bag, and fastened it under her skirt. When she left the shop, the son

---

[‡]See also An Ancient History of Poisons in the Introduction.
[§]He escaped in 1773 and went into hiding for a time, but then continued with his dissolute way of life.

followed her and then alerted the authorities who arrested her. At her trial for theft, it was revealed that she had taken around 11 pounds (5 kg) of cantharides in total and that this was worth 9 shillings a pound, adding up to a total of £4 19s, which in modern terms would be equivalent to around £500. Elizabeth was found guilty of what was a then a hanging offence, but her life was spared because she was pregnant, a condition confirmed by a group of midwives specially called to the court to examine her.

The 1850s saw another more serious trial involving cantharidin, as we shall see.

13.3 MEDICAL USES

When a weak solution of cantharidin is applied to the skin, it immediately forms a blister. This outcome was put to use in order to remove skin blemishes such as warts. A wart is first pared away as far as possible. Then, cantharidin solution is applied to it and to the surrounding area, although this should extend only a fraction of an inch beyond the wart itself. This results in being able to lift the wart off the skin and hence remove it. However, this could only be done under medical supervision, and the solution so used consisted of 0.7% cantharidin in acetone. Such solutions are still available, if no longer used medically, and they often contain camphor or pine oil to give them a kind of medical aroma. One manufacturer reported selling 150 litres a year of such medication just before the US Food and Drug Administration (FDA) de-listed it in 1962. This decision was made because of the failure of manufacturers to reveal how they extracted it and how they had tested it for safety.

Nevertheless, some practitioners continue to use it as way to remove warts, tattoos, and the papules of the virus molluscum, which appear, generally on children, as pink pustules. A more dilute solution was even used as a cosmetic to give skin a rosy glow and to give hair a golden tint.

Cantharidin is absorbed by the fatty epidermal cells of the skin and the alimentary canal, which extends from the mouth to the anus, and there it interferes with two key enzymes, phosphatase 2A and serine protease. Their role is to block chemicals which are released by cells undergoing breakdown and which would otherwise attack normal healthy cells. These chemicals are now free to attack the peptide bonds of proteins, which are essential for the structure of cells. As these bonds break, cells then detach themselves and serum oozes to occupy the space and so blisters form. Such blisters do not result in permanent scarring – at least, if the affected person can survive their encounter with cantharidin.

If someone gets cantharidin on their skin, then the treatment is to wash the area with soap and water several times, or to wipe with an organic solvent like acetone, alcohol, or ether, but blistering may still occur up to six hours later.

13.4 CANTHARIDIN AS A POISONOUS CHEMICAL

The **LD_{50}** of cantharidin for humans is 0.5 mg per kg body weight so that for a typical 70 kg person, a dose of 35 mg might well be fatal, and 70 mg would certainly be so. Even a dose as low as 10 mg has been known to kill, and yet others have survived doses of 50 mg. Whatever the dose, it will certainly cause a serious illness. When taken by mouth, cantharidin will cause a burning sensation of the lips, mouth, and pharynx within a few minutes.

Cantharidin is rapidly absorbed by the body, which quickly recognises that this chemical is not needed and immediately the kidneys begin to excrete it. A non-fatal dose will disappear from the body within three to four days. However, the kidneys can also fall victim to the toxin if the dose is large enough, and they will respond by swelling and may soon fail altogether.

The response to cantharidin depends on the dose, of course, and ranges from mild discomfort to extreme pain and death. Too much cantharidin and the most obvious symptoms are the vomiting of blood, the passing of bloody diarrhoea, convulsions, drop in blood pressure, leading ultimately to death. Other symptoms are dark urine containing blood which is painful to pass, groin pain, vaginal bleeding, and rectal bleeding. Near-fatal doses will involve the vomiting of blood and an irregular heartbeat.

Cantharidin's toxicity is such that a fisherman who had some on his fingers – he was intending to catch fish using it – pricked his thumb with a fishing hook, and this caused him to suck it, which in turn, transferred enough of the toxin into his mouth to affect him fatally. In 1996, four students at Temple University in

Philadelphia were admitted to a local hospital after they had experimented with cantharidin, adding it to Kool-Aid powder from which drinks are made. They were lucky and all survived.

There is no known antidote for cantharidin poisoning. The victim should be encouraged to drink lots of water, but not to vomit, and maybe given activated charcoal to absorb the toxin. Liquid paraffin mineral oil can be used to speed evacuation of the bowels. Fatty substances should be avoided because they will aid the action of the poison. Once it is clear the patient will survive, then calcium and magnesium supplements need to be given for several weeks. Painkillers will almost certainly be needed.

In 1954, the *British Medical Journal* published an article giving the methods by which cantharidin might be identified: a sample had to be extracted and its melting point measured – this should be 212 °C; a little was to be applied to an arm and this should result in characteristic blistering; and an X-ray diffraction pattern should be recorded for the crystalline material. The blister test for cantharidin could be performed by rubbing material taken from inside the body of the deceased against a patch of shaved skin of a rabbit. Today, none of this is necessary because this drug can easily be confirmed by **HPLC–MS** (high performance liquid chromatography–mass spectrometry) analysis.

13.5 SPANISH FLY PIE, ACCIDENTAL POISONING, 1850

Celebrity status may flatter well-known actors, but it can also attract the attention of someone whom they do not wish to know. In 1850, Martha Sharp was just such a someone, and James Elphinstone, who was a principal actor of the Pavilion Theatre in the East End of London, did not want to know her. He was a married man and she appeared to be a prostitute.

On 23 July, Ms Sharp was having no luck in contacting the comedy actor so she devised a plan to meet him. Elphinstone was already aware of Ms Sharp and deliberately avoided her outside the theatre and when she followed him to the local pub, the King's Head, where he often went after his performance. She had tried to speak to the actor on several occasions when she had been to the theatre with her friend, Mrs Holborough, and her friend had managed to attract the actor's attention and had a conversation with him. However, when she suggested he should meet Ms Sharp, he declined the invitation.

Frustrated by her several attempts to meet the actor, Ms Sharp decided to take things further, and she baked him a jam tart, at the bottom of which she put some Spanish fly. She handed this to a waiter at the King's Head pub and asked him to give this to the actor when he came there that evening after his performance.

In fact, the waiter handed it to Thomas King, who was Mr Elphinstone's dresser, asking him to give it to the actor. Elphinstone told King to keep it for himself, and so he took it home and gave it to his wife Charlotte. That evening she ate only a little of it, but the following morning she consumed almost all the rest of it, although not that part in the centre that she later described as "the green stuff" which lay under the layer of jam. Not long after, she was violently sick and this brought back up most of the tart along with quantities of blood.

Dr Henry Cornelius, who lived in Whitechapel High Street, was sent for and dealt with her as best he could. Then he asked what had caused her illness and she told him what she had eaten. He took away the remains of the jam tart and found it contained several bits that he could identify as bits from beetles that were a golden-green colour. He estimated that there were about five grains of insect material in the tart (325 mg).

Martha Sharp was arrested and sent for trial at the Old Bailey, when Dr Cornelius gave evidence. He told the court that he was aware of the effects of cantharides because he had attended lectures about its use and misuse by the famous toxicologist Dr Taylor. His treatment of Charlotte helped her recover, although at the time of the trial he said she was still in a delicate state of health. When questioned by the counsel for the defence regarding the use of cantharides by "women of this class," the doctor said that he was aware that they believed it would increase the passion of their customers.

The court finally had to deal with what was rather a difficult issue in that while Ms Sharp had intended to injure the actor, she did not intend to injure his dresser's wife, so there was no case to answer. The jury agreed and she was found not guilty.

13.6 ACCIDENTAL MURDER, LONDON, 1738

On 18 December 1738, a young apprentice, Michael Dunn, died under mysterious circumstances at his brother's home in Westminster. He told those who attended him that he had not been well after drinking some coffee which he had been given on the morning of 15 December by Catherine Demay. It appeared she had poisoned him with cantharidin. She was arrested and accused of his murder and brought to trial at the Old Bailey on 17 January 1739.

Michael Dunn was a journeyman[¶] for William Charlton, a barber and wig-maker, and he lived with his master at Cock Court off Ludgate Hill in London. On the morning in question, Catherine, who was a live-in servant, had called down the stairs from her attic room saying she had made a pot of coffee and wondered if Michael would like some, but she got no response because he was out on an errand. Charlton's son went up to her room and brought the coffee pot downstairs, where he put it by the fire to keep warm. When Michael returned, he drank three cups of coffee from it and ate some buttered toast. On finishing his third cup, Charlton noted that it contained what appeared to be green fragments of something he thought were green tea-leaves, although they seemed to be of rather an odd shape. Three hours later, Michael began to feel rather strange.

What was later said in court about his strange condition could not be written in the record on the grounds of its being indecent. This rather suggests that his strange condition was a prolonged erection. "An uncommon ailment" was how one witness described it when he came to the shop and spoke with Michael later that day. The customer told the court that Michael's symptom "was in the lower part of his body *etc. etc.*"[‖] and that he was blaming green particles in the coffee which he had drunk. He later told another witness, who saw him on the morning of his death, two days later, that he had drunk coffee which had a great many green spangles in it.

However, it had not been until the morning of Sunday that Michael complained to his master of feeling very ill and

---

[¶]Today, we'd call him a gofer.

[‖]The court record refers to what is clearly a prolonged erection in various ways such as "an alteration in himself," "a disorder," and "the effects of the poison."

thought he might benefit from walking to his brother's home in Westminster, about a mile away. He set off, but he was never to return to the rather aptly named Cock Court because he died at his brother's house the following day. Michael's brother, Brian, was out when Michael arrived and so he waited on a seat outside the pub next door, complaining of feeling very ill. When Brian came home, he helped Michael into his house and gave him some ale and treacle before putting him to bed. By now he was sweating profusely and was clearly in agony. At midnight, Michael told his brother he was going to die and he appeared to be choking. By morning, he was very thirsty but would not drink, saying he was burning up inside.

Brian went to get a local apothecary, and they returned to find Michael in the pub, where he was clinging to the bar post and asking the lady of the house to help him. They got him back to bed, and it was then that Michael said to his brother that "the d*mned b**ch Demay" (as written in the court records) gave him the coffee which poisoned him. Soon after he said this, he died. Demay was arrested and charged with Michael's murder.

Two hours before Michael died, a Mr Tagg, who was described as a surgeon, was sent for, and he found Michael in great pain and said that three people had to hold him down so Tagg could bleed him. Another surgeon was sent for but it was too late, and Michael died before the surgeon arrived to bleed him further. Then a physician was sent for, a Dr Connel, who performed a kind of autopsy which revealed that the heart and stomach were vastly inflamed and his stomach contained a powder which Connel could not identify. Mr Tagg was also there at the post-mortem and described the stomach as vesicated (in other words, blistered) and with part of it loose, as if there had been a large blister which had burst. Another blister appeared to be full of dirty mucous. He also said that Michael had had large evacuations, both upwards and downwards. However, there was no evidence of "the foul disease." Tagg was asked how long it might take for cantharidin to have an effect, and he replied it would depend on the quantity taken but, at maximum, this would be 12 hours. The defence questioned whether it would be possible for someone who had been given the drug on Friday morning to have survived to Monday, to which Tagg replied that it would depend on the contents of the stomach and he had known of men who had taken as long as eight days to respond to Spanish fly.

At the trial, Demay accused the brother of being malicious; "You'd spend £100 to hang me," she cried out in court. Brian admitted he'd said as much because he felt she was guilty.

Demay's defence was that Michael had been with prostitutes and had "got the clap."** However, several witnesses at the trial, such as his aunt who attended him during his final hours, confronted the sick man with the evidence of his erect penis, which she assumed had resulted from his becoming infected by a prostitute. He denied ever having lain with a woman. Another witness, Ann Dewsbury, who saw him on the Monday morning also asked him, when she was alone with him, if his "disorder" had been caused by his being "among lewd women." Again, he denied this was the cause of his condition and blamed the coffee.

In court, an apothecary was called to give evidence and he was asked what he thought was the cause of death. He said that only cantharides could explain Michael's condition, and that he was well versed in the effects of this drug, having been an apothecary for 22 years. The counsel for the defence then asked if this drug might have been taken to treat "the French disease"[††] and was told that this was sometimes the case.

The defence now called several witnesses. Their aim was to show that Michael had been a bit of a lad-about-town and that was the cause of his strange condition and death. First, they called Mary Winnel to the stand and she attested that she had breakfasted with Catherine Demay that Friday morning and had been asked whether she preferred tea or coffee, to which she answered coffee. Catherine then went to the shop on the corner and bought half an ounce of coffee and this she brewed. They drank it together and ate oat cake. She said that at no time did Catherine add anything to the coffee. She attested to the good character of the prisoner.

A Jane Marsh then was called and she refuted the claims that Michael had been unwell on the Saturday, saying he had been quite merry that evening.

**Gonorrhoea.
[††]Syphilis.

A James Gilstrop gave evidence that early in December he had been out walking with Michael, who said he was suffering a "foul distemper" and had been taking pills. He also said that the two men had on occasions "gone up the stairs" with girls. At one time, Michael had lodged in the same house as himself in Covent Garden and had showed him some "pills for fun" and bragged about the effect they had. Another witness told of Michael having a phial of liquor which he'd got from a doctor at the "other end of town," which by implication, meant it was of a sexual nature. A John Thompson also said that Michael had a box of pills and a phial in November and that they were able to cure a "young man of the aforesaid malady," which the court record does not specify. A man called Phipp, who was employed by a brandy merchant, testified that he and Michael had been out on the town and ended up in Drury Lane, where they had been with two girls, after which Michael told Phipp that he had given his girl the clap.

A Samuel Farthing, who was employed by a chemist in Ludgate Hill, said that Michael had come into the shop with a note asking for elixir *antivenerum unciauna* (meaning anti-venereal disease, one ounce), which he said was for a friend, but he'd refused to serve him.

A Mr William Morgan then gave an account of the effects of cantharides, saying that this would affect a person instantly it was taken and questioned the findings of those who had carried out the perfunctory autopsy. Others then appeared as character witnesses for Catherine.

The result of all this was a verdict of "not guilty."

So what do we make of all this? What caused the court recorder, and some of the witnesses, no little embarrassment was describing the chief symptom which Michael was displaying on the Saturday and Sunday of the weekend when he died. No doubt he had taken some Spanish fly as a desiccated and ground-up specimen of the green beetle variety. Clearly he had taken too much and then tried to hide what was happening to him. When questioned, he blamed the coffee, but had he added the Spanish fly himself? Obviously, someone had to be blamed for what had happened but there was no proof that Catherine, or her friend, was responsible. Or maybe they had done it as a joke, knowing the effect it would have on Michael.

13.7 UNINTENTIONAL MURDERS, LONDON, 1954

Arthur Kendrick Ford was a 44-year-old married man with two young children, and he worked as an office manager for a wholesale pharmaceutical firm based on the Euston Road in London. His story begins on a homeward-bound ship bringing Ford back from war service in the Far East. It was on that journey he learnt of the effects of Spanish fly and how this would affect both men and women, resulting in increased sexual urges. However, Ford did not know how this wonderful agent could be obtained, but he didn't forget about it until one day, and quite by chance, he discovered that Spanish fly was really a chemical called cantharidin.

Ford's office staff consisted of 22 women, most of whom were typists, and four men. One of the women had caught his eye and he flirted with her. She was 27-year-old Betty Grant, and he later claimed that she had responded to his advances to the extent that the two of them stayed behind after work and had sex. Indeed, he knew Betty outside of working hours because she lived near the Fords and had acted as a babysitter for them on several occasions.

Early in April, Ford had taken a phone call from a customer asking if the firm could supply him with the drug cantharidin, which Ford had never heard of, but he said he would make enquiries. He went to the stock room and spoke to the man in charge, a Mr Richard Lushington, who was a trained chemist. He learned that cantharidin was the chemical name for Spanish fly and he was shown a bottle. He now lied to Lushington, saying that a neighbour of his wanted some of the drug to pep up the rabbits that he was using for breeding purposes. Lushington said that the drug was highly dangerous and refused to give him any on account of its being a Schedule I poison. However, Ford did note where the bottle of cantharidin was kept and, during a lunch break, he went to the stock room and put about 40 grains of the powder (around 2.5 grams) into an envelope and took it back to his desk. When Mr Lushington later checked the contents of the bottle, after it was clear that its contents had been misused, he was able to calculate how much Ford had taken.

Monday 26 April was the fateful day. Ford went out of the office at lunchtime to a local sweetshop and bought four ounces (110 grams) of coconut ice, which amounted to eight pieces. These were cubes made of desiccated coconut and sugar; half of each cube being pink and half white. While he was alone in his office, Ford took out his envelope of cantharidin and pushed a small amount of the drug into two of the cubes using the tips of a pair of scissors.

When the staff had returned from their lunch break, Ford walked around the office offering the coconut ice to some of the staff, although he chose the piece he handed to Betty Grant. It is thought that he may have intended eating the other piece of doctored coconut ice himself, but he gave it to June Malins, for reasons that were never made clear. She was 19 years old and was very good-looking; indeed, she had taken part in seaside beauty competitions the previous year and been crowned queen at one of them. Ford also gave the remaining coconut ice pieces to other girls and they saw him eat a piece as well.

An hour later, June said she was feeling ill and she was taken to the first-aid room, where she vomited and complained of severe stomach pains. It was assumed to be just a reaction to something she had eaten at lunchtime and she was left to recover. Half an hour later, Betty too went to the first-aid room with the same symptoms. She then decided to take a taxi home and reached there at 6 pm. She vomited in the taxi several times. Her local doctor was called and saw her at 7:15 pm, when he immediately arranged for her to be admitted to St James's Hospital in Balham, south London. She arrived there at 9:30 pm, by which time she was vomiting blood. Her stomach was washed out and she was given an injection of morphine, and her condition appeared to be improving so she was admitted to one of the wards. Soon afterwards, she vomited almost half a pint (280 mL) of blood and her pulse was erratic. She died at 7 am the following day. A post-mortem was performed later that morning.

Meanwhile, June Malins had been admitted to University College Hospital in a collapsed state. She was first given a drink of sodium bicarbonate solution to treat the terrible indigestion she was complaining of. Then she started vomiting blood-stained mucous and had a blister at the side of her mouth. Her pain became much worse, and she had difficulty speaking and swallowing. She was now given morphine and put on an intravenous drip, and for 14 hours she remained sedated but then her pulse began to rise and, because she had passed no urine by the following morning, a catheter was inserted at 9 am, but only a little blood-stained urine was extracted. Other medication was added to her intravenous drip but to no avail, and she died at 4:40 pm that day. One of the doctors treating her suspected she had been poisoned with cantharidin.

Ford too went round to University College Hospital on that fateful day, saying he was suffering from some peculiar blisters on his face, although he displayed no sign of poisoning. He no doubt had transferred a little cantharidin to his face while he was putting it into the coconut sweets.

Autopsies were carried out on the two women and it was obvious that they had died of cantharidin poisoning. When Ford was told, his first response was "Oh, dear God, what an awful thing I have done. Why didn't somebody tell me!" and he then confessed to the doctors what his role had been in causing their deaths.

Why Ford also gave a piece of the coconut ice to June Malins is unclear, but it may be that he originally intended that piece of coconut ice was for himself and somehow had inadvertently given it to June. Or did he fancy his chances with her as well as with Betty?

When detectives went to examine the firm's offices, they found traces of crystals on Ford's desk and a pair of scissors with smear marks on its tips. These were sent to the Metropolitan Police Laboratory for analysis and were identified as cantharidin crystals and cantharidin mixed with coconut ice, respectively. Samples of vomit and various organs from Betty Grant and June Malins were also sent for analysis and attempts were made to extract cantharidin from these, but this proved difficult in the case of the vomit, but was possible from the organs, which were liquidised and extracted with chloroform. When this was evaporated, crystals were obtained and X-ray analysis showed they were cantharidin. The blister test was also used to show that it was indeed cantharidin.

Ford was arrested and brought to trial at the Old Bailey and charged with manslaughter. He was found guilty and sentenced to five years in prison. In a newspaper account after the trial – they paid for his defence for exclusive rights to publish his story – Ford maintained that he had fallen in love with Betty Grant. However, he had not taken this so far as to have a sexual relationship with her as he claimed at the trial, as the forensic pathologist who testified said in court: she had died a virgin.

13.8 MURDER, MELBOURNE, AUSTRALIA, 1960

When 23-year-old Marion Sapwell was asked to be a bridesmaid at the wedding of a friend, she was delighted. She was originally from New Zealand and had gone to Australia to complete her nursing studies at Melbourne's Box Hill Hospital, where she had made several friends. When she was qualified, she then moved to a permanent job in Wangaratta District Hospital. This served a town of around 18 000 inhabitants in the state of Victoria and was about 150 miles from the capital Melbourne.

On Friday 22 July 1960, she drove to Melbourne with the object of spending some time with the bride-to-be before the wedding, and she stayed with friends, Mr and Mrs Etienne Mehari. However, Mr Mehari had designs on Marion, and for breakfast that Saturday morning, he put some cantharidin in her coffee, which she said tasted a little funny when she drank it. Marion began to feel ill soon after breakfast, and indeed was so ill that a doctor was called and he admitted her to Box Hill Hospital later that morning. Within 24 hours, she was dead.

Dr James McNamara of the Forensic Medicine Department of the University of Melbourne was asked to examine her body. He was also an assistant medical officer of the City Coroner. He noted parchment-like areas of skin on her body and blistering of her internal organs. He knew immediately that she had been poisoned with cantharidin, and this was confirmed by Lynn Turner, the deputy medical-legal chemist for the state of Victoria. It was later estimated that she had consumed quite a lot more than 10 mg of the poison.

Mr Mehari was 30 years old and lived with his wife and their young son in a flat over the pharmacy which his wife managed.

He was employed by a firm of wholesale druggists. The Meharis had been introduced to Marion by a mutual friend of theirs who lived in New Zealand, and she brought a letter of introduction for them when she arrived in Melbourne.

Who had given Ms Sapwell the cantharidin? The police suspected Mr Mehari, but why would he give her so much cantharidin as to cause her serious harm? When he was interviewed by detectives a few days after her death, he said he had dosed Marion with a view to having a sexual encounter with her but did not want his wife to know. On the evening of 26 July 1960, he was charged with murder.

Mehari's trial began in November of that year, and straight away he denied ever telling detectives that he had given Marion cantharidin and accused the detectives of framing him and even hitting him. He also said that the detectives threatened to prosecute his wife if he did not admit to what he had done, and they told him to think of his young son if this were to happen.

The defence was that Ms Sapwell had died from an overdose of self-administered cantharidin. As a nurse, she could have had access to this drug and thought it might have an effect on herself. Did she want to sex herself up? Or maybe she took it to procure an abortion. The defence counsel also called their own expert, Frank Shaw, Professor of Pharmacology at the University of Melbourne, who disagreed that the death was due to cantharidin poisoning, pointing out that the "parchment-like" abrasions on Ms Sapwell's skin were not at all like the blisters that cantharidin produced. He even showed the court two such blisters on his own arm and challenged the prosecution witness, Ms Turner, to say whether cantharidin caused them. She was unable to say for certain whether this was so.

No other tests had been conducted on samples taken from Ms Sapwell's body, so the forensic evidence was rather sparse. Nor had police been able to find any cantharidin in the Mehari's flat or at his place of work, and the pharmacy where his wife worked did not have any either. On 30 November, the last day of the trial, the judge told the jury that there was no proof that the poison which killed Ms Sapwell was cantharidin, but whether it was or not, the jury found Mr Mehari guilty of manslaughter, and on 3 December, he was sentenced to five years' imprisonment.

Whatever was in Ms Sapwell's coffee that morning was quick-acting and had a detrimental effect on the lining of her stomach and led her to suffer the usual symptoms of cantharidin poisoning. The police suspected Mehari of being responsible for her death, although why he put cantharidin in her coffee was never explained.

CHAPTER 14

Hemlock at the End of It All

A word in **bold** *indicates that further information can be found in the Glossary. Only the first time the word appears in a chapter will it be so indicated.*

Double, double, toil and trouble
Fire burn and caldron bubble
....
Scale of dragon, tooth of wolf
Witches' mummy, maw and gulf
Of the ravin'd salt-sea shark
Root of hemlock, digg'd i' the dark . . .

So chanted the witches in Act 4 Scene 1 of Shakespeare's play *Macbeth*, and he certainly knew that one of the things they were adding to the pot was very dangerous. That was the root of the hemlock plant – and it continues to exert its spell, as we shall see.

Hemlock is one of the plants most likely to cause poisoning because it is easily confused with other plants that are edible: its fruits resemble those of aniseed; the leaves are like parsley; and the roots resemble parsnips. What might warn someone that it is not the real thing is its faint odour of urine.

There are several varieties of hemlock, of which two are the most common and likely to cause poisoning if consumed. They

More Molecules of Murder
By John Emsley

Published by the Royal Society of Chemistry, www.rsc.org

are *Conium maculatum*, more commonly called spotted hemlock, which derives its deadly potency from the molecule **coniine**, and *Cicuta maculata*, more commonly called water hemlock, which gets its toxicity from the equally deadly molecule **cicutoxin**. The two molecules are very different chemically. It is coniine and its effects which will mainly be discussed in this chapter because spotted hemlock is the plant that people generally encounter, at least in the UK. It is mostly to be found growing in waste land or in woodland. Water hemlock, as its name implies, is found in swampy regions.

Hemlock plants have everyday names, which vary from country to country, and there are several varieties of this plant, such as *Cicuta bulbifera*, *Cicuta virosa*, and *Cicuta occidentalis*. And there are an equal number of local names for hemlock, such as bunk in Canada and cicuta in South America. In the UK, the *Cicuta virosa* (Figure 14.1) hemlock is also known as cowbane. However, it is not the variety of plants that we are concerned with here, but the toxic chemicals which they produce.

Figure 14.1 Poisonous water hemlock root (*Cicuta virosa*), which provides cicutoxin; first isolated in 1914. © Hein Nouwens/Shutterstock.

14.1 CONIINE

Coniine is not the only **alkaloid** that *Conium maculatum* produces but it is by far the deadliest.[†] Coniine is a chiral molecule, which means that it can exist in two forms that are mirror images of each other, and sometimes referred to as right-handed (*dextra* or *d*) and left-handed (*laevo* or *l*) isomers; they are also called enantiomers or optical isomers. Many molecules in Nature are of this kind and, while the two isomers look almost the same to the untrained eye, they can behave very differently from each other. In some cases, only one of the isomers can trigger a response at a receptor while the other isomer cannot. This is particularly marked with fragrance molecules. With coniine, the effect is to make one chiral isomer, the *l* form, deadlier than the *d* isomer. When coniine is prepared in the lab, it consists of equal amounts of both isomers.

The hemlock plant has lots of leaves and is one of the first plants to emerge in spring. It grows to be about three feet tall and has small white flowers which then produce the tiny fruits which look like those of the caraway. All parts of the plant contain coniine and the leaves can have between 1–2% of the toxin which, when pure, is a colourless oil that boils at around 170 °C. It contains coniine in all its parts but especially in the roots and seeds, and yet it is said that larks and quails are able to eat the seeds and not be affected by them, although the flesh of these birds then becomes dangerous to consume. In Italy, there were cases reported in the 1970s and 1980s of people being poisoned by hemlock from eating birds which had fed on hemlock seeds. Four of those affected died of renal failure. Thrushes are also said to be able to eat the seeds with impunity. The larvae of some moths, such as the silver-ground carpet moths, can also use hemlock leaves as a source of food.

Coniine was first extracted from hemlock by the chemist L. Gieseke in 1827. Its chemical formula analysed as $C_8H_{17}N$ but its molecular structure was not to be known for another 50 years, until it was deduced by the great German chemist August Wilhelm Hofmann (1818–1892). The molecule has quite a simple structure consisting of a six-membered ring of five carbons plus one

---

[†]The others are *N*-methyl-coniine, γ-coniceine, conhydrine, pseudoconhydrine, conhydrinone, and *N*-methyl-pseudoconhydrine.

nitrogen (piperidine) with a short chain of three carbons (a propyl group) attached to one of the carbons next to the nitrogen. In 1886, another German chemist, Albert Landenburg (1842–1911), showed it was possible to make coniine in the lab. Simple as the molecule is in chemical terms, it is an insidious poison which can easily infiltrate the body and begin its deadly progress.

Coniine targets the central nervous system, briefly stimulating it but then slowly closing it down, thereby causing muscle paralysis and so stopping the working of the heart and the lungs. Depending on the dose, the only outward signs are likely to be copious saliva production and possibly vomiting. There will be a rapid increase in heartbeats and high blood pressure and dilation of the pupils. Slowly, the muscles will become paralysed, starting from the feet and moving upwards.

The best treatment for someone who has ingested hemlock is immediate pumping out and washing of the stomach, followed by the administration of activated charcoal, after which the medical care will be to treat symptoms and provide general support. The victim will also need artificial respiration while he or she copes with the coniine that is in their body and while the metabolic processes seek to remove it completely. This will generally take two to three days, but full recovery will then result.

Coniine can even be absorbed through the skin, and sometimes, the leaves have been used as a plaster to cure pain. The analgesic power was the main reason why dried leaves from the plant were once part of the herbalist's repertoire and taken as such. Then it became available in the form of the salt it forms with hydrobromic acid (HBr) and, as such, it was listed in pharmacopeias. It was to be taken in doses of about 1 mg. The recommended maximum dose was 15 mg per day. Traditional herbalists still offered it as the powdered dried fruits of the plant, or it was possible to make a tea from fresh hemlock leaves and that was made with around one gram of the leaves in 100 millilitres of water.

Coniine is rather like curare in its mode of action – see Chapter 11. It also resembles **nicotine** and acts in a similar way. Its site of action in the body is at the synapse between neurons, where **acetylcholine** is a key molecular messenger released by the end of one nerve and which crosses the gap between them and triggers another nerve by locking on to its receptors. Coniine blocks the receptors, ultimately stopping the beating of the heart. It slowly blocks the receptors of the central nervous system, starting from the feet and working upwards through the body.

As with so many natural products, there were always those who were prepared to experiment with hemlock, hoping to discover that, in small, non-lethal doses its active component would have some healing powers. In Medieval times, it was claimed that hemlock was a cure for the bite of a mad dog. Hemlock juice was once considered a treatment for teething in children, for epilepsy, bronchitis and whooping cough, for which it was thought to make breathing easier by relaxing the muscles of the windpipe (trachea). It was also used in the treatment of tuberculosis, tumours, and pains in the joints.

We have seen with several of the toxins discussed in this book how potentially deadly poisons were often used medically, and hemlock was no exception. However, we now know that for no illness could it be a cure. Nevertheless, coniine was once part of the *British Pharmacopoeia*, where it was listed as a sedative and with the ability to control spasms. It was even suggested as an antidote to **strychnine** poisoning. The pharmacopeia also referred to the dried leaves and the fruit of hemlock as possible medicaments, but ideally, an extract of hemlock was to be given in the form of a tincture, which is a solution in alcohol.

In fact, there was no good reason to prescribe coniine for any condition or illness, and it was removed from the *British Pharmacopoeia* in 1934.

14.2 CICUTOXIN

Cicutoxin was never listed in pharmacopeias. Nor did herbal practitioners realise that water hemlock had a very different molecule hidden in its foliage and roots which could poison those who ate them. However, they did know to treat it with the same respect as the more common spotted hemlock. As with that other type of hemlock, the roots of water hemlock offered the same possibilities of being mistaken for food plants such as carrot, parsnip or ginseng. The leaves are also toxic but not to the same extent.

Eating any part of water hemlock was always dangerous and led to accidental and deliberate poisoning over the centuries. For example, in 1962 there were a reported 78 incidences of this happening to people, of whom 33 died. We will look at a more recent case below. There was also the rare case of a young boy who made a whistle out of the stem of the plant and was poisoned by it. Cicutoxin can penetrate the skin, as a family discovered when they were using the leaves of the plant to relieve themselves of excessive itching (pruritus) and thereby became ill.

Cicutoxin disrupts the central nervous system by interfering with the mechanism by which GABA (gamma-aminobutyric acid) operates, and it blocks the sites that GABA needs to activate. These control the essential movement of chloride and potassium ions across the membrane of key components of the immune system, such as the lymphocytes in the blood, and as it deactivates more and more of the lymphocytes, then the body responds and the symptoms appear: vomiting, excessive salivation, sweating, widening of the pupils, to be followed by the more serious reactions of convulsions, seizures, and coma. Respiratory failure is likely and will lead to death. Even if this does not occur, then it is more than likely that the kidneys will fail, and eventually, this might cause death. Cicutoxin is only slowly excreted by the body and it may take up to seven days for it to be removed. The effect that cicutoxin has on the lymphocytes led, in 1986, to its investigation as a possible treatment for leukaemia.

The treatment of someone suspected of cicutoxin poisoning is to wash out the stomach and administer activated charcoal, and this should be done within 30 minutes of water hemlock being eaten. Artificial respiration is also needed to prevent suffocation.

Cicutoxin was first isolated in 1914 by C. A. Jacobsen, who recognised it as the poisonous principle in water hemlock, and he was able to isolate it and analyse it as $C_{17}H_{22}O_2$. But what was its structure? This curious molecule has a chain of carbon atoms along which there are three double bonds and two triple bonds. Synthesising it in the lab did not occur until 1955 and then only in a tiny amount. Its actual molecular structure was only deduced in 1999.

14.3 HEMLOCK AS A POISON

This section generally refers to the more common spotted hemlock.

Hemlock is deadly. The **LD_{50}** for animals is estimated to be 3 mg kg^{-1} for cows, 15 mg kg^{-1} for horses, and 45 mg kg^{-1} for sheep. However, for humans the fatal dose is said to be 100–130 mg, and this will produce death within three hours, and it will progress along the lines described for Socrates – see Section 14.5 below. With a fatal dose, the most obvious symptoms are those affecting the peripheral nervous system, although the central nervous system is what closes the body down permanently. A dose of 3 mg of coniine will produce symptoms, although barely noticeable, whereas a dose of 150 mg should be fatal but it was reported to have no effect in one individual. For most individuals, a dose of 50 mg might well be dangerous and, certainly, a dose of 100 mg is likely to kill. This is the kind of dose you would get if you ate around seven leaves of hemlock.

Rarely has hemlock been used to murder because it is too slow to act and thereby the victim may survive if medical help arrives in time. However, it was the poison used by one of the characters in an Agatha Christie book: *Five Little Pigs*, which appeared in 1942. Although that tells of a deliberate poisoning with hemlock, it does not appear to have been based on an actual murder. However, there are several cases of hemlock killing by accident.

For example, in 1845, there was the unfortunate death of Duncan Gow. He was a tailor in Edinburgh, and he was struggling to bring up a family and was happy to make economies for himself when it came to eating; so, naturally, he was pleased when his girls brought him a sandwich for lunch one day that they had prepared, which contained parsley as one of its components – or so they thought and he assumed was the case. Unfortunately, it was hemlock leaves, which the children had gathered while out walking that morning. Later that afternoon, Gow's legs became numb and he was unable to walk. Then, slowly, the numb feeling began to affect the rest of his body until he was incapable of any motion, although he was able to talk. Slowly, he died. The Scottish pathologist John Bennett (1812–1875) performed a post-mortem and confirmed the cause of death as being due to hemlock.

More recently, there was another accidental, but tragic, poisoning which took place in Maine, USA. In October 1992, two brothers decided to search for wild ginseng in the local woods. The younger of the two, who was 23 years old, pulled up a variety of plants that he found in a swampy area, and one in particular looked very promising. It had a solid root that he thought might be ginseng so he took several bites from it, chewed them, and swallowed them. His elder 39-year-old brother also ate a small piece of the root.

Half an hour later, the younger brother began to feel very odd in his legs and could hardly walk; he sat down and then began to have convulsions. His brother rang the emergency services, who quickly arrived. They saw that he was in a bad way. He was blue in the face, salivating, with his pupils dilated. They gave him artificial respiration and rushed him to the nearest hospital. The doctors there used gastric lavage to wash out the contents of his stomach and administered activated charcoal. His heart was now beating erratically and he was having seizures. Four times he needed to resuscitating, but ultimately, he was beyond saving and he died. That occurred about three hours after he had eaten the hemlock root.

His older brother responded positively to the same treatment and, although he became delirious and had seizures, his life was saved and, after a few days of intensive care, he was safely discharged from hospital. The root the brothers had eaten was identified as *Cicuta maculata*, although it was first suspected that they had been affected by cicutoxin and that they had consumed some root of another water hemlock, *Cicuta virosa*. This also produces cicutoxin. However, **HPLC–MS** (high performance liquid chromatography–mass spectrometry) tests on samples from the younger brother's liver, blood, and gut contents showed cicutoxin not to be present, although it was said that it might have been the cause of death and had degraded in the samples before analysis.

14.4 POSSIBLE ASSASSINATION, BABYLON, 323 BC

Alexander the Great died in 323 BC at the height of his power. What killed him? He was struck down with something which caused a fever and confusion in his speech. There have been suggestions that he was poisoned and there were plant-derived ones which were known at the time, such as strychnine and hemlock. However, strychnine is not consistent with the symptoms he displayed in his final week of life.

At the beginning of June 323 BC, Alexander became ill when he was staying in the royal palace at Babylon, a city which had surrendered to him two years previously. However, it was not just a fever that he was suffering from, as it first appeared, and he slowly got worse and died on 11 June. He was still only 32 years old and is now regarded as one of the greatest generals in history. Although he was actually King Alexander III of Macedon, in Northern Greece, he would forever be known as Alexander the Great on account of his remarkable military achievements, having conquered much of the ancient world from Egypt to Afghanistan. He had founded more than 20 towns and cities that were named after him – some of which are still so named today.

His death at so early an age has always raised suspicions that he was poisoned, and several suggestions have been made as to the agent which was used. Since his body and tomb have never been found, it is impossible to rule out mineral poisons such as arsenic. Poisons derived from natural plants would not survive in what remains of a body after such a long period, even if it were discovered.

A recent article published in the journal *Clinical Toxicology* in 2014, by Leo J. Schep *et al.* of the University of Otago, Dunedin, New Zealand, suggested that the symptoms Alexander displayed during the 11 days of his final illness were consistent with being poisoned by an extract from the plant *Veratrum album*, commonly known as hellebore.

However, that theory was questioned in a letter to the journal by Christopher Wiart of the University of Nottingham who pointed out that this plant could not have been used because it did not grow anywhere near Babylon and is only to be found at high altitudes. Wiart suggests that a more likely poison would have been hemlock, which was a popular poison used in ancient Greece to dispatch opponents. Wiart also quotes from a letter written to Alexander by the physician Androcyde, who advised Alexander to drink wine as an antidote to hemlock, this being assumed to be the cause of his condition.

The symptoms displayed by Alexander during his dying days, of a high temperature, general debility, confused speech, and weakness in his legs, are consistent with hemlock being the poison and this being given to him on more than one occasion. He might also have been given an extract from water hemlock. If Alexander's death were to be attributed to hemlock, then it must have been given to him in more than one dose because he took almost 10 days to die.

Normally, the effects of hemlock appear within hours, assuming a fatal dose has been given, as we discover from another well-documented death by poison from Greek history: the execution of Socrates. We know for certain that this was due to common hemlock.

14.5 EXECUTION, GREECE, 399 BC

Probably the most famous poisoning in history was that of the 70-year-old philosopher Socrates in 399 BC. He had been accused of corrupting the youth of Athens with his anti-democratic views, and his trial was before a jury of 500 citizens, in which three accusers presented the case against him. Socrates was given an equal amount of time to defend himself. A vote of the jury was then taken but it went against him: 280 voted him guilty, 220 not guilty.

Next, the jury had to decide on his punishment. His accusers wanted him executed, but the court said he could decide his own punishment, hoping that he would chose exile, which would have been acceptable. Instead, the old man suggested he be fined a derisory amount, but the jury was not in the mood to be mocked and he was sentenced to death. The method of execution in Athens was by means of poison, and the poison was hemlock.

Socrates' most famous student was Plato, who was in his mid-to-late 20s when all this happened.[‡] He later wrote about his mentor's death, basing it on the accounts of other students who actually witnessed it, and there were reputed to be 14 of them. The poison was prepared by one of the guards and delivered to Socrates ready-mixed in a cup, which he then drank. As he did

[‡]His date of birth is uncertain but was between 423 and 429 BC.

so, he prayed to the gods and hoped that his life beyond the grave would be a happy one. He was told to walk around once he had drained the cup and to do this until his legs went numb, when he could lie down. His students burst into tears at what was happening, only to be told to control themselves. "I sent the women away to stop them creating such a scene. What is wrong with you?" said Socrates. "I should like to die in silence" and he covered his head with a cloth.

As he lay on a couch, the executioner squeezed the philosopher's feet and asked him if he could feel anything. Socrates replied that he could not. Next, he squeezed his thighs and got the same answer. Quite quickly, the numbness reached his waist and clearly it would soon reach his heart. However, he just had time to make one last request and that was to tell his pupil Crito that he owed Asclepius a cock and to pay the debt. Crito agreed. Then Socrates covered his face again. When Crito next asked Socrates if there was anything else he wanted him to do, he got no reply. The executioner eventually removed the cloth and examined Socrates' eyes, which were open but which did not respond to his touch. So he closed them and closed his mouth as well. The philosopher was dead, but he and the poison that killed him were to be forever famous.

Literature

G. Bell, *The Poison Principle; a Memoir of Family Secrets and Literary Poisonings*, Macmillan, Australia, 2002.
D. Blum, *The Poisoner's Handbook*, Penguin Books, New York, 2010.
W. H. Brook, *The Case of the Poisonous Socks; Tales from Chemistry*, Royal Society of Chemistry, Cambridge, 2011.
J. Buckingham, *Bitter Nemesis; the Intimate History of Strychnine*, CRC Press, Boca Raton, Florida, 2008.
T. T. Buhk, *Michigan's Strychnine Saint; the Curious Case of Mrs Mary McKnight*, The History Press, Charleston, 2014.
I. Butler, *Murderer's England*, Robert Hale, London, 1973.
C. Cobb, M. L. Fetterolf and J. G. Goldsmith, *Crime Scene; Chemistry for the Armchair Sleuth*, Prometheus Books, Amherst, 2007.
P. Cooper, *Poisoning by Drugs and Chemicals, Plants and Animals*, Alchemist Publications, London, 3rd edn, 1974.
Fundamental Toxicology for Chemists, ed. J. H. Duffus and H. G. J. Worth, Royal Society of Chemistry, London, 1996.
More Chemistry and Crime, ed. S. M. Gerber and R. Saferstein, American Chemical Society, Washington DC, 1997.
J. Glaister, *The Power of Poison*, Christopher Johnson, London, 1954.
The Greek Herbal of Dioscorides, ed. R. T. Gunther, translated by John Gooyer, Oxford University Press, Oxford, 1934.

More Molecules of Murder
By John Emsley

Published by the Royal Society of Chemistry, www.rsc.org

J. Emsley, *Elements of Murder*, Oxford University Press, Oxford, 2005.

J. Emsley, *Molecules of Murder*, Royal Society of Chemistry, Cambridge, 2008.

C. Evans, *The Casebook of Forensic Detection; How Science Solved 100 of the World's Most Baffling Crimes*, John Wiley & Sons Inc., New York, 1996.

C. Evans, *Murder 2; The Second Casebook of Forensic Detection*, Wiley, Hoboken New Jersey, 2004.

S. Feldman, *Poison Arrows: the amazing story of how Prozac and anaesthetics were developed from deadly jungle poison darts*, Metro Publishing, London, 2005.

P. Frank and M. A. Ottoboni, *The Dose Makes the Poison*, Wiley, Hoboken, 3rd edn, 2011.

More Chemistry and Crime: From Marsh Arsenic Test to DNA Profile, ed. S. M. Gerber and R. Saferstein, American Chemical Society, Washington DC, 1997.

J. Goulter, *No Verdict, New Zealand's Hung Jury Crisis*, Random House, Auckland, 1997.

K. Harkup, *A is for Arsenic; the Poisons of Agatha Christie*, Bloomsbury, London, 2015.

J. M. Hightower, *Diagnosis: Mercury; Money, Politics & Poison*, Island Press, Washington, 2009.

K. Hollington, *How to Kill; The Definitive History of the Assassin*, Arrow Books, London, 2007.

M. B. Jacobs, *The Analytical Chemistry of Industrial Poisons, Hazards, and Solvents*, Interscience, New York, 2nd edn, 1949.

D. Janes, *The Poisoned Partridge; the Strange Death of Lieutenant Chevis*, The History Press, Stroud, 2013.

R. M. Julien, *A Primer of Drug Action*, Freeman & Co, New York, 10th edn, 2004.

B. H. Kaye, *Science and the Detective*, VCH, Weinheim, 1995.

M. Kelleher and C. L. Kelleher, *Murder Most Rare: the Female Serial Killer*, Dell Publishing, New York, 1998.

S. Kind, *The Sceptical Witness*, Hodology Ltd, The Forensic Science Society, Harrogate, 1999.

C. D. Klaassen, *Casarett & Doull's Toxicology; the Basic Science of Poisons*, McGraw Hill, New York, 6th edn, 2001.

C. Lavery, *The Black Widower*, Mainstream Publishing, Edinburgh, 2012.

A. J. Lax, *Toxin; the Cunning of Bacterial Poisons*, Oxford University Press, Oxford, 2005.

The Pharmaceutical Codex, ed. W. Lund, The Pharmaceutical Press, London, 12th edn, 1994.

J. Mann *et al.*, *Natural Products,* Addison Wesley Longman Ltd, Harlow UK, 1996.

J. Mann, *Murder, Magic and Medicine,* revised edn, Oxford University Press, Oxford, 2000.

T. McLaughlin, *The Coward's Weapon*, Robert Hale, London, 1980.

A. C. Moffat, M. D. Osselton, B. Widdop and J. Watts, *Clarke's Analysis of Drugs and Poisons*, Universal Free e-Book Store, 4th edn, 2016.

H. Montgomery Hyde, *Crime Has its Heroes*, Constable, London, 1976.

B. Morrissey, *When Women Kill; Questions of Agency and Subjectivity*, Routledge, Abingdon Oxfordshire, 2003.

A. Motion, *Wainewright the Poisoner,* Faber and Faber, London, 2000.

R. Odell, *The Mammoth Book of Bizarre Crimes; Incredible Real-Life Murders*, Constable & Robinson, London, 2010.

The Merck Index; an Encyclopaedia of Chemicals, Drugs, and Biologicals, ed. M. J. O'Neil, Royal Society of Chemistry, Cambridge, 15th edn, 2013.

M. A. Ottoboni, *The Dose Makes the Poison*, Van Nostrand Reinhold, New York, 2nd edn, 1991.

C. E. Overton, *Studies of Narcosis*, Chapman and Hall, London, 1991.

C. A. Pasternak, *The Molecules Within Us*, Plenum Trade, New York, 1998.

P. Paul, *Murder Under the Microscope*, Macdonald, London, 1990.

C. J. Polson and R. N. Tattersall, *Clinical Toxicology*, EUP, London, 1965.

E. Rentoul and H. Smith, *Glaister's Medical Jurisprudence and Toxicology*, 13th edn.

R. Rowland, *Poisoner in the Dock*, Arco, London, 1960.

Science Against Crime, various contributors, Marshall Cavendish, 1982.

Taylor's Principles and Practice of Medical Jurisprudence, ed. K. Simpson, 12th edn, Churchill, London, 1965, vol. II.

N. Sly, *Murder by Poison; a Casebook of Historic British Murders*, The History Press, Stroud Gloucestershire, 2009.

S. D. Stevens and A. Klarner, *Deadly Doses: a Writer's Guide to Poiosons*, Writer's Digest Books, Cincinnati, Ohio, 1900.

A. Stolman and C. P. Stewart, 'The absorption, distribution, and excretion of poisons' in *Progress in Chemical Toxicology*, 1965, vol. 2, p. 141.

T. Stone and G. Darlington, *Pills, Potions and Poisons*, Oxford University Press, Oxford, 2000.

L. Stratmann, *The Secret Poisoner; a Century of Murder*, Yale University Press, New Haven, 2016.

Handbook of Analytical Toxicology, ed. I. Sunshine, Chemical Rubber Co., Cleveland, Ohio, 1969.

C. J. S. Thompson, *Poisons and Poisoners*, Harold Shaylor, London, 1931.

J. Thorwald, *Proof of Poison*, Thames & Hudson, London, 1966.

J. A. Timbrell, *Introduction to Toxicology*, Taylor & Francis, London, 1989.

J. A. Timbrell, *The Poison Paradox; Chemicals as Friends and Foes*, Oxford University Press, 2005.

J. H. Trestrail III, *Criminal Poisoning; Investigational Guide for Law Enforcement, Toxicologists, Forensic Scientist, and Attorneys*, Humana Press, Totowa New Jersey, 2nd edn, 2007.

Martindale; the Extra Pharmacopoeia, ed. E. Wade, The Pharmaceutical Press, London, 27th edn, 1977.

Molecules of Death, ed. R. H. Waring, G. B. Steventon, S. C. Mitchell, Imperial College Press, London, 2nd edn, 2007.

K. Watson, *Poisoned Lives*, Hambledon, London, 2004.

Crime Scene to Court, ed. P. C. White, Royal Society of Chemistry, Cambridge, 2004.

C. Wilson, *Written in Blood; A History of Forensic Detection*, Equation Press, Wellingborough Northamptonshire, 1989.

C. Wilson and P. Pitman, *Encyclopaedia of Murder*, Arthur Barker, London, 1961.

G. Winger, F. G. Hofman and J. H. Woods, *A Handbook of Drug and Alcohol Abuse*, Oxford University Press, New York, 1992.

R. A. Witthaus, *Manual of Toxicology*, William Wood, New York, 1911.

J. C. Wood, *The Most Remarkable Woman in England; Poison, Celebrity and the Trials of Beatrice Pace*, Manchester University Press, 2012.

A. C. Wootton, *Chronicles of Pharmacy*, Milford House, Boston, 1910 (republished 1971).

Glossary

Acetylcholine

Acetylcholine is a positively charged molecule and so needs a negative ion such as chloride or bromide to balance it. Its chemical name is 2-(acetoxy)-*N*,*N*,*N*-trimethylethanaminium and its chemical formula is $C_7H_{16}NO_2$. It acts as a neurotransmitter (nerve messenger) and is vital in activating muscles, including those of the heart. Within the brain, it is needed to process information. After acetylcholine has does its job, it is removed by the enzyme acetylcholinesterase, which breaks it down into choline $(CH_3)_3NCH_2CH_2OH$ and acetic acid CH_3CO_2H. Any chemical which interferes with acetylcholine or deactivates this enzyme will disrupt vital functions of the body.

Aconitine

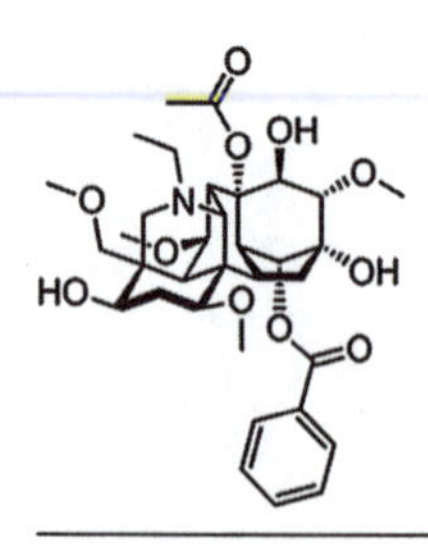

More Molecules of Murder
By John Emsley

Published by the Royal Society of Chemistry, www.rsc.org

The chemical name is (1α,3α,6α,14α,15α,16β)-8-acetoxy-20-ethyl-3,13,15-trihydroxy-1,6,16-trimethoxy-4-(methoxymethyl)aconitan-14-yl-benzoate and its chemical formula is $C_{34}H_{47}NO_{11}$. Aconitine is poisonous, with an **LD_{50}** for mice of 1.8 mg kg^{-1}. Aconitine is soluble in chloroform, less soluble in alcohol, and almost insoluble in water, with only one gram dissolving in 4.5 litres of water, whereas this amount will dissolve in 40 mL of alcohol, or 3 mL of chloroform.

Acrylamide

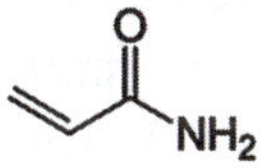

O
NH$_2$

The chemical name for acrylamide is prop-2-enamide and its chemical formula is C_3H_5NO. As depicted, it is referred to as a monomer and this melts at 85 °C. When heated to high temperatures, it polymerises. The monomer is very soluble in water, to the extent that a litre of water will dissolve 2 kg. It is also very soluble in alcohol.

Alkaloid
This is a class of chemical compounds produced by plants, fungi, and even animals. The alkaloids all contain a neutral nitrogen atom which is what makes them react with acids and so makes them appear to be like alkalis, hence the name. Many have been used in traditional medicines, such as quinine for malaria and morphine as a painkiller. They are also used in modern medicine and some have been modified chemically to produce new drugs. They can be highly toxic, such as **strychnine**.

Alum
This is a double sulfate. Its chemical name is potassium aluminium sulfate and its chemical formula is $KAl(SO_4)_2$, although it generally occurs as the hydrate $KAl(SO_4)_2 \cdot 12H_2O$. It is used in water purification, tanning leather, dyeing, and it is even a component of baking powder. Styptic pencils, used to stop the bleeding caused by small cuts when shaving, contain alum.

Amitriptyline

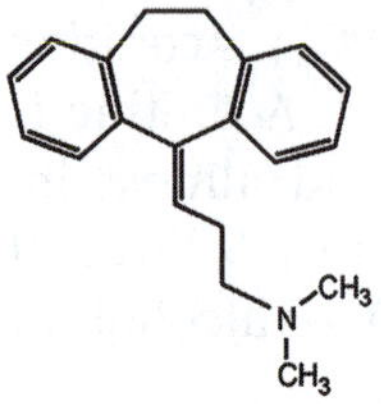

The chemical name is 3-(10,11-dihydro-4*H*-dibenzo[*a*,*d*]cycloheptene-5-ylidene)-*N*,*N*-dimethyl-1-propanamine and its chemical formula is $C_{20}H_{22}N$. There are various trade names for this medication, such as Elavil and Endep. It is an antidepressant, and it can relieve long-term pain associated with arthritis. It also relaxes muscles and induces sleep.

Atropine

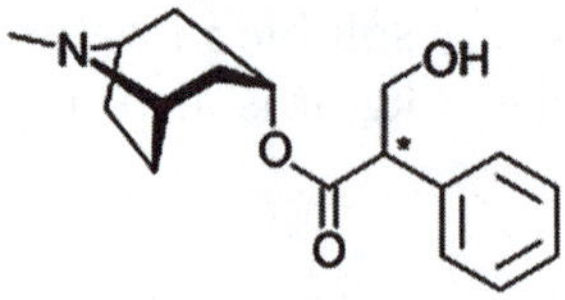

The chemical name for this **alkaloid** is α-(hydroxymethyl)-benzeneacetic acid (-*endo*)-8-methyl-8-azabicyclo[3.2.1]oct-3-yl ester and its chemical formula is $C_{17}H_{23}NO_3$. It can be obtained as long crystals from acetone solution and it melts at 114–116 °C. It takes 455 mL of water to dissolve one gram of atropine at room temperature. Atropine is prescribed when someone is suffering from pesticide poisoning and to treat conditions such as slow heartbeat, as well as being used in surgery to supress saliva.

Beta particle (negative)
This is an electron emitted by a neutron in the nucleus of an atom, whereby it undergoes radioactive decay and is thereby converted to a proton. The particle is usually written as β^-. When this happens, the atomic number of the nucleus increases by +1, thereby converting that atom to an atom of the element one unit higher in the periodic listing. (There are also beta particles with positive charge, known as positrons and written as β^+.)

Caffeine

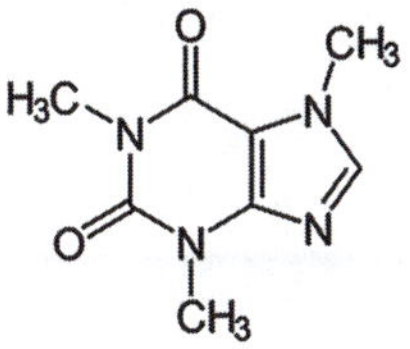

The chemical name for this is 3,7-dihydro-1,3,7-trimethylpurine-2,6-dione and its chemical formula is $C_8H_{10}N_4O_2$. It forms crystals which sublime when heated to 160–165 °C. It is soluble in most solvents. The $\mathbf{LD_{50}}$ for mice is 127 $mg\,kg^{-1}$.

Cantharidin

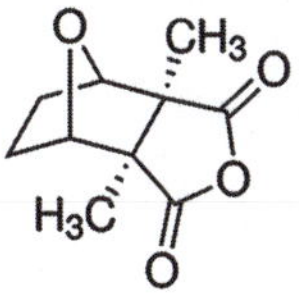

Cantharidin belongs to the terpenoid group and its chemical name is (1*R*,2*S*,6*R*,7*S*)-2,6-dimethyl-4,10-dioxatricylo[5.2.1.$0^{2,6}$]decane-3,5-dione and its chemical formula is $C_{10}H_{12}O_4$. The molecule consists of three interconnected rings and an anhydride (–CO–O–CO–) unit.

In contact with skin, it acts as a vicious blistering agent. It is commonly known as Spanish fly and was for centuries used to boost sexual activity.

Cicutoxin

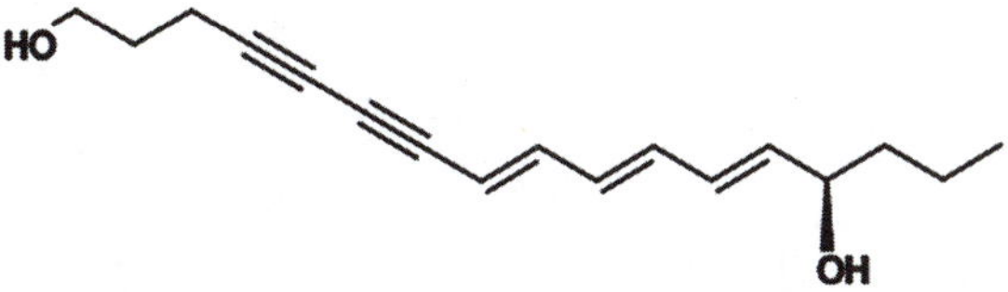

The chemical name is (8*E*,10*E*,12*E*,14*R*)-8,10,12-heptadecatrien-3,6-diin-1,14-diol and the chemical formula is $C_{17}H_{22}O_2$. Plants form it as just the chiral isomer whereas when prepared in the lab, it is a racemic mixture. This is the natural toxin that is mainly present in the root of the water hemlock, and it causes death by disrupting the central nervous system.

Coniine

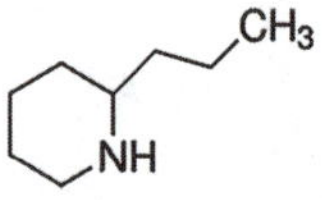

The chemical name is (2*S*)-2-propylpiperidine and the chemical formula is $C_8H_{17}N$. The molecule exists as a mixture of both left and right-handed structures, sometimes referred to as (±)-coniine. This is the toxin that is present in common hemlock plants.

Cysteine

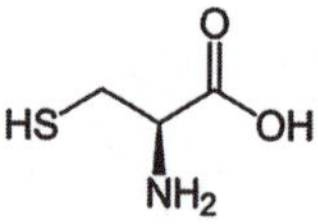

The chemical name for this is 2-amino-3-(2-amino-2-carboxy-ethyl)disulfanyl-propanoic acid and its chemical formula is $C_3H_7NO_2S$. This is an amino acid and its SH group is often involved in enzyme chemistry, and as such, cysteine is quite important in metabolism and is regarded as an important part of the diet. However, it is not absolutely essential, and is a part of most proteins.

Diazepam

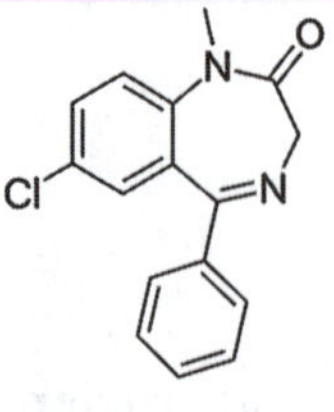

The chemical name for this molecule is 7-chloro-1-methyl-5-phenyl-1,3-dihydro-2*H*-1,4-benzodiazepin-2-one and its chemical formula is $C_{16}H_{13}ClN_2O$. It is more popularly known as Valium, which is prescribed to treat depression and similar conditions. Chemically, it comes as pale yellow crystals which melt at 131–134 °C. It is virtually insoluble in water but soluble in organic solvents.

Difenacoum

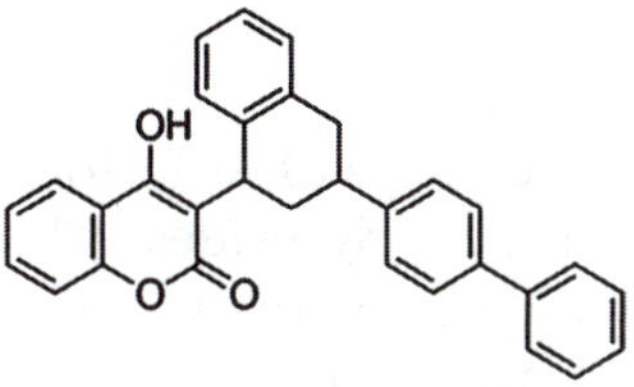

The chemical name is 2-hydroxy-3-[3-(4-phenylphenyl)-1-tetralinyl]-4-chromenone and its chemical formula is $C_{31}H_{24}O_3$. Its crystals melt at 215–217 °C. Its **LD$_{50}$** for rats is 2.5 mg kg^{-1}, hence its use as a rodenticide. It can be prescribed medically as an anticoagulant.

Digitoxin

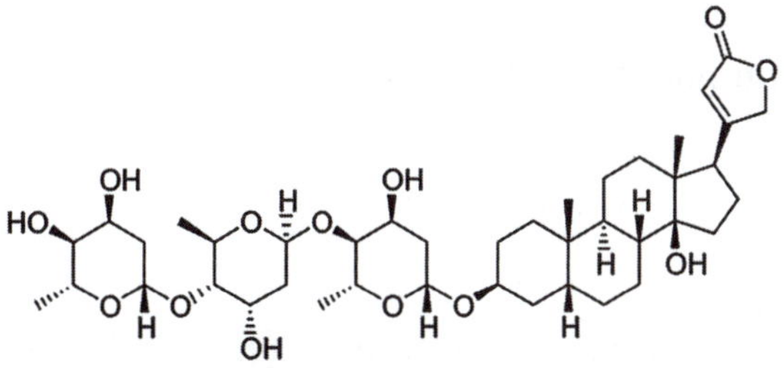

Digitoxin consists of a short carbohydrate chain with a lactone ring at one end. It is one of a group called glycosides. The chemical name is 3-[*O*-2,6-dideoxy-β-D-*ribo*-(1→4)-*O*-2,6-dideoxy-dideoxyhexopyranosyl-β-D-*ribo*-(1→4)-2.6-dideoxy-β-D-*ribo*-hexopyranosyl]-14-hydroxycard-20(22)-enoloide and its chemical formula is $C_{41}H_{64}O_{13}$. When pure, its melting point is 256–257 °C. It is practically insoluble in water, but soluble in chloroform and alcohol. The **LD$_{50}$** for mice is 5 mg kg^{-1}. It is no longer prescribed medically because **digoxin** performs better, although neither is much used.

Digoxin

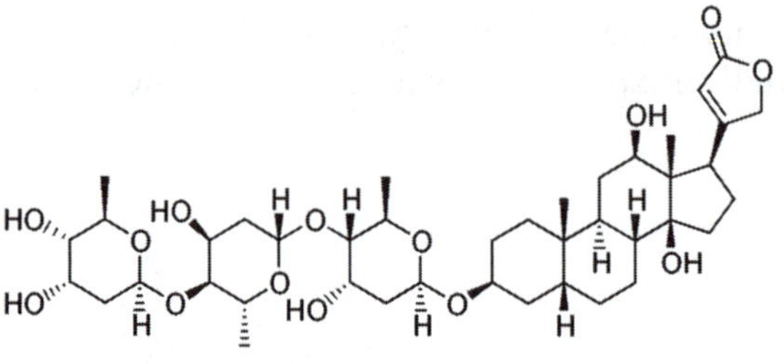

Digoxin consists of a short carbohydrate chain with a lactone ring at one end. It is one of a group called glycosides. The chemical name for digoxin is (3β,5β,12β)-3-{[2,6-dideoxy-β-D-*ribo*-hexopyranosyl-(1→;4)-2,6-dideoxy-β-D-*ribo*-hexopyranosyl-(1→4)-2,6-dideoxy-β-D-*ribo*-hexopyranosyl]oxy}-12,14-dihydroxycard-20(22)-enolide and its chemical formula is $C_{41}H_{64}O_{14}$. Digoxin is almost insoluble in water, slightly soluble in ethanol, and freely soluble in chloroform and methanol. On heating, it begins to decompose at 230 °C. The **LD_{50}** for mice is 18 $mg\,kg^{-1}$. Although once used to treat heart disease, it has been superseded by other drugs.

Dimercaptopropanesulfonate (DMPS)

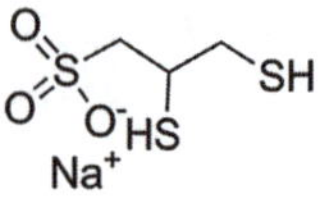

The name generally refers to the sodium salt, as shown, and its chemical name is sodium (2*S*)-2,3-bis(sulfanyl)propane-1-sulfonate and its chemical formula is $C_3H_7NaO_3S_3$. It is sometimes prescribed to treat heavy metal poisoning.

Epsom salts

The chemical name for this is magnesium sulfate and its chemical formula is $MgSO_4$, although it generally comes with seven water molecules as the hydrate $MgSO_4 \cdot 7H_2O$. The single water hydrate $MgSO_4 \cdot H_2O$ is the mineral kieserite, which is mined to the extent of two million tonnes a year or more, most of which is used in fertilizers.

Eserine – see **physostigmine.**

Ethyl acetate

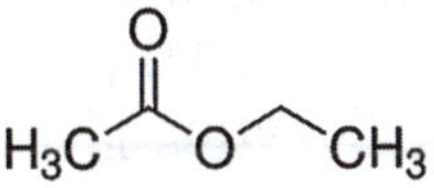

or $CH_3–C(O)–O–CH_2CH_3$

The chemical name for this molecule is acetic acid ethyl ester and its chemical formula is $C_4H_8O_2$. It is a colourless liquid with a fruity odour and its boiling point is 77 °C. It is used as an organic solvent and in personal products, such as nail varnish remover.

Ethylene glycol

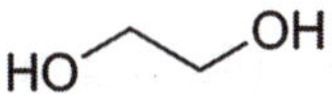

or $HO–CH_2–CH_2–OH$

The chemical name is ethane-1,2-diol and its chemical formula is $C_2H_6O_2$. This is a flammable liquid which boils at 14 °C. It is made by the catalytic oxidation of ethylene, $H_2C{=}CH_2$, with air. Reacted with water, it converts to mono-ethylene glycol (MEG), di-ethylene glycol (DEG), and tri-ethylene glycol (TEG). These are used to de-ice aircraft but may damage the environment.

Fomepizole

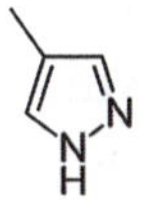

The chemical name for this is 4-methylpyrazole and its chemical formula is $C_4H_6N_2$. It is also known as Antizol and is an antidote to poisoning by methanol and ethylene glycol. Fomepizole is a low melting solid (melting point 25 °C).

Gelsemine

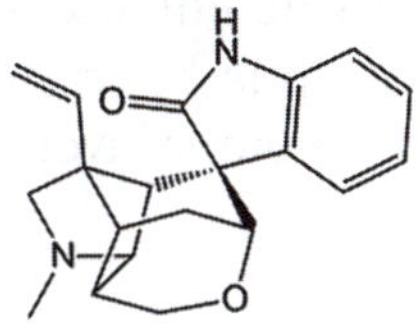

The chemical name for this **alkaloid** is (3*S*,3a*S*,4*S*,5*R*,8*S*,8a*S*,9*R*)-3-ethenyl-1-methyl-2,3,3*a*,7,8,8*a*-hexahydro-1*H*,5*H*-spiro[3,8,5-(ethane[1,1,2]triyl)oxepino[4,5-*b*]pyrrole-4,3′-indol]-2′(1′*H*)-one and its chemical formula is $C_{20}H_{22}N_2O_2$. When pure, it is a white crystalline powder with a melting point of 178 °C. The $\mathbf{LD_{50}}$ for mice is 300 mg kg^{-1}. Its name comes from the flowering plant *Gelsemium sempervirens.*

Glycerophosphate, sodium salt

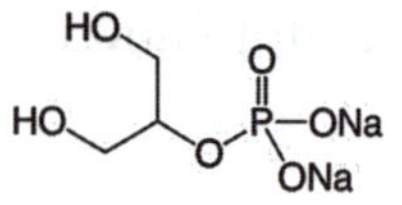

or $HOCH_2CH(OH)CH_2OPO_3Na_2$

The chemical name is disodium 2,3-dihydroxypropyl phosphate and its chemical formula is $C_3H_3O_6PNa_2$. Glycerophosphate is needed for bone (calcium phosphate) formation and it is also involved in muscle cells.

Glycolaldehyde

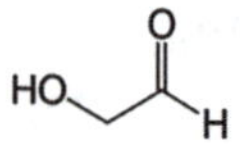

or $HO–CH_2–CHO$

The chemical name is either 2-hydroxyacetaldehyde or 2-hydroxyethanal and its chemical formula is $C_2H_4O_2$. It is a solid which melts at 97 °C. This is a molecule that exists in space and is thought to be one of the molecules that were building blocks for early life-forms.

Glycolic acid

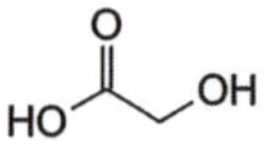

or $HOC(O)CH_2OH$

The chemical name is 2-hydroxyethanoic acid and its chemical formula is $C_2H_4O_3$. It melts at 75 °C. This is an alpha-hydroxy acid and is used in personal care products to remove outer layers of dead skin.

HPLC–MS (High performance liquid chromatography in tandem with mass spectroscopy)

This is the most powerful analytical technique, capable of detecting and identifying the minutest amounts of a chemical even when this is mixed with many other things. The sample is first separated into its multiple components by means of liquid chromatography, and each component is then identified by mass spectrometry. It is used in forensic detection and in many other fields of research, namely pharmaceuticals, agrochemicals, food contamination, and other industries.

Isopropyl alcohol

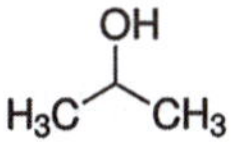

OH
H_3C CH_3 or $CH_3CH(OH)CH_3$

The chemical name is propan-2-ol and it has the chemical formula C_3H_8O. This is a useful solvent, mixing easily with water, and it boils at 80 °C. The liquid is colourless and will burn if ignited. It is used industrially as a solvent and in personal care products, such as rubbing alcohol and in hand sanitisers. It is also used as a petrol additive.

LD_{50}

This term means lethal dose 50% and is shorthand for the amount of a chemical that will be lethal to half of the animals to which it is given. When the amount so calculated is converted to the amount that would similarly affect humans then it is assumed that the average person weighs 70 kg (11 stone), although most adults today weigh more than this.

Nicotine

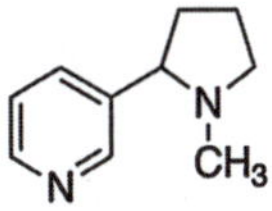

N
N
CH_3

The chemical name is 3-[(2*S*)-1-methyl-2-pyrrolidinyl]pyridine and its chemical formula is $C_{10}H_{14}N_2$. This is the mildly addictive agent that smoking tobacco provides, as do the electronic versions. Nicotine stimulates certain sensors in the brain, although in larger amounts it is a deadly poison.

Nitrocellulose

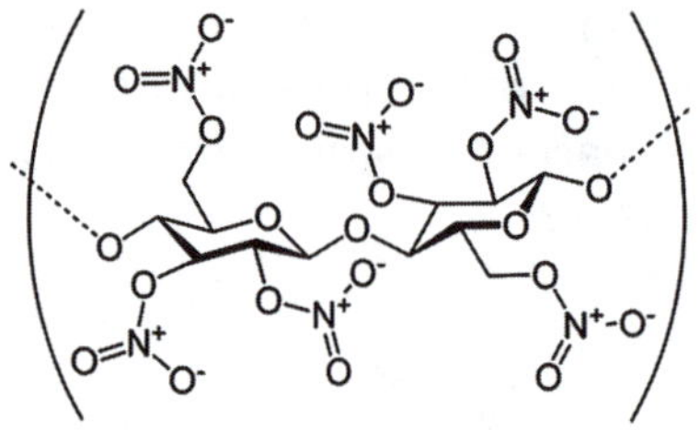

This is the compound formed by the reaction of nitric acid and cellulose, and it has been employed as gun cotton and was used to make the transparent plastic film for motion pictures.

Oxalic acid

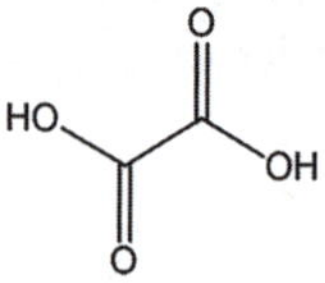

or HO–(O)C–C(O)–OH

The chemical name is ethanedioic acid and the chemical formula is $C_2H_2O_4$. It is a colourless, crystalline solid which, when crystallised from water, comes as the dihydrate: $HO_2CCO_2H \cdot 2H_2O$ and, as such, will effloresce (lose water slowly) and then it becomes a powder. It is used as a cleaning agent to remove limescale and microbes.

Palmitic acid

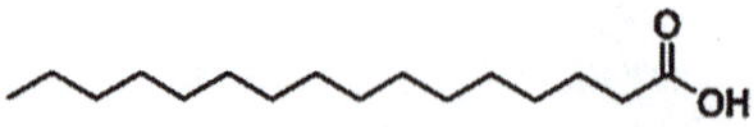

or $CH_3(CH_2)_{14}COOH$

The chemical name is hexadecanoic acid and its chemical formula is $C_{16}H_{32}O_2$. Palmitic acid is a saturated acid and is the most common fatty acid that animals and plants produce, and it is especially abundant in the oil of the palm tree from which it gets its name. It is a solid which melts at 63 °C.

Physostigmine

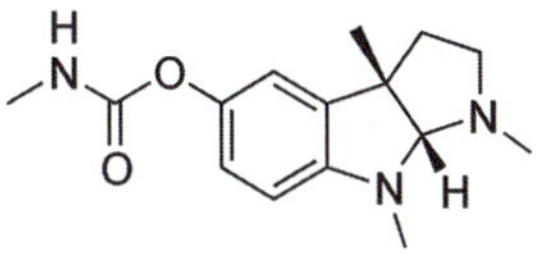

The chemical name is (3a*R*,8a*S*)-1,3*a*,8-trimethyl-1*H*,2*H*,3*H*,3a*H*,8*H*,8a*H*-pyrrolo[2,3-b]indol-5-yl *N*-methylcarbamate and its chemical formula is $C_{15}H_{21}N_3O_2$. This **alkaloid** is also known as serine, the name deriving from the name of the Calabar bean of West Africa, which is called éséré. It has an **LD_{50}** of 3 mg kg^{-1} for mice. It interferes with the **acetylcholine** involved in the central nervous system and has been used medically to treat glaucoma and for gastric disorders.

Phytonadione

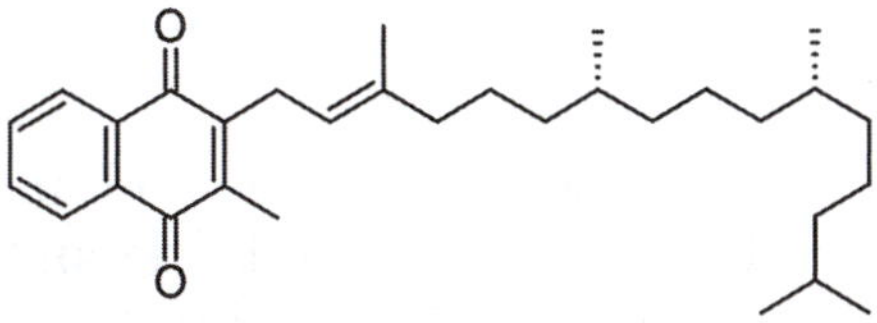

The chemical name is 2-methyl-3-[(2*E*,7*R*,11*R*)-3,7,11,15-tetramethylhexadec-2-en-1-yl]-1,4-dihydronaphthalene-1,4-dione 2-methyl-3-[(2*E*)-3,7,11,15-tetramethylhexadec-2-en-1-yl]naphthoquinone and its chemical formula is $C_{31}H_{46}O_2$. It is also known as vitamin K and is essential for bone formation.

Scopoletin

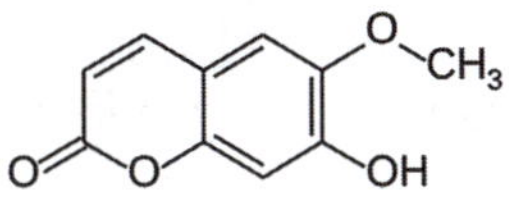

The chemical name is 7-hydroxy-6-methoxy-2*H*-1-benzopyran-2-one and its chemical formula is $C_{10}H_8O_4$. It occurs in the roots of several plants, such as chicory and stinging nettle. It can control blood pressure and has a role in alternative medicine.

Strychnine

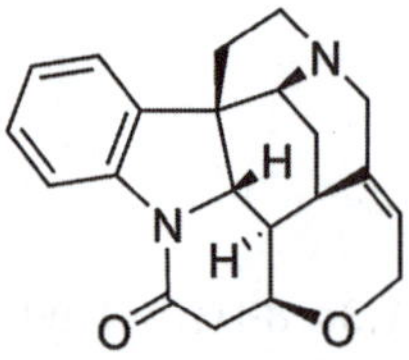

The chemical name is 4*a*,5,5*a*,7,8,13*a*,15,15*a*,15*b*,16-decahydro-2*H*-4,6-methanoindolo[3,2,1-*ij*]oxepino[2,3,4-*de*]pyrrolo[2,3-*h*]-quinoline-14-one and its chemical formula is $C_{21}H_{22}N_2O_2$. As the pure chemical, it forms colourless, needle-like crystals. It is almost insoluble in water but its salts are soluble.

Sulfanilamide

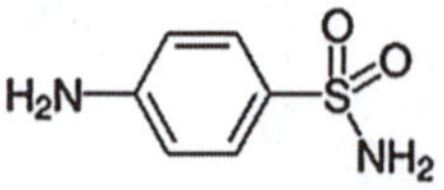

The chemical name is 4-aminobenzenesulfonamide and its chemical formula is $C_6H_8N_2O_2S$. It was the first antibiotic and widely used until it was supplanted by more effective ones.

Temazepam

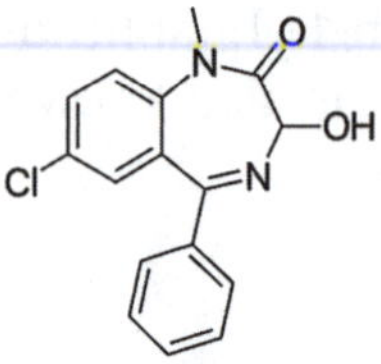

The chemical name is 7-chloro-1,3-dihydro-3-hydroxy-1-methyl-5-phenyl-1,4-benzodiazepin-2-one and its chemical formula is

$C_{16}H_{30}ClN_2O_2$. Temazepam is a white, crystalline substance, very slightly soluble in water and sparingly soluble in alcohol. It can be prescribed to treat insomnia.

Tetramethylenedisulfotetramine (aka tetramine)

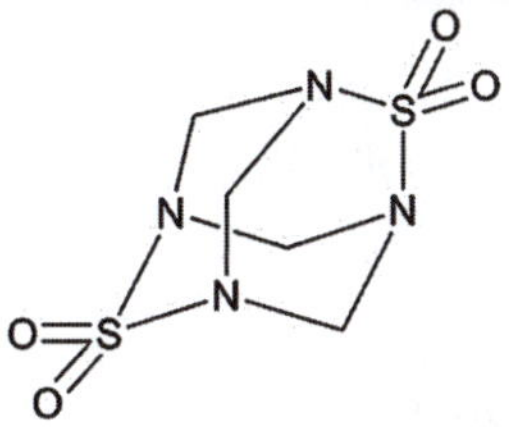

The chemical name is 2,6-dithia-1,3,5,7-tetraazatricyclo[3.3.1.$1^{3,7}$]-decane 2,2,6,6-tetraoxide, and its chemical formula is $C_4H_8N_4O_4S_2$. This is a highly toxic chemical, previously used as rat poison but now banned worldwide.

Subject Index